T0341200

Healthcare Affordability

Continuous Improvement Series

Series Editors:
Elizabeth A. Cudney and Tina Kanti Agustiady

Design for Six Sigma: A Practical Approach through Innovation
Elizabeth A. Cudney and Tina Kanti Agustiady

Affordability: Integrating Value, Customer, and Cost for
Continuous Improvement
Paul Walter Odomirok, Sr.

Continuous Improvement, Probability, and Statistics: Using
Creative Hands-On Techniques
William Hooper

Transforming Organizations: One Process at a Time
Kathryn A. LeRoy

Statistical Process Control: A Pragmatic Approach
Stephen Mundwiller

For more information about this series, please visit: https://www.crcpress.
com/Continuous-Improvement-Series/book-series/CONIMPSER

Healthcare Affordability
Motivate People, Improve Processes, and Increase Performance

Paul Walter Odomirok, Sr.

CRC Press
Taylor & Francis Group
Boca Raton London New York

CRC Press is an imprint of the
Taylor & Francis Group, an **informa** business

Contents

Preface

Shortly after the Theory of Affordability materialized in 2007 and the foundational text, *Affordability: Integrating Value, Customer and Cost for Continuous Improvement* was released in December 2016, it became apparent to me that although affordability could be achieved for any specific product or service, its scope can range across entire enterprises and markets. Affordability provides a new and unique, triple-lens view through which to see, observe, and examine an enterprise, and create a focus to aim and balance value, customer, and cost. Inherent is the capability to optimize (1) customer requirements and the delivery of value, (2) expenses and resources for value delivery, and (3) create competitive pricing that generates substantial revenue for maintaining the organization's sustainability that increases demand and stimulates growth.

Affordability for the greater good of society can be targeted at enterprises such as healthcare, education, government, banking, retail, manufacturing, as well as others. This book is focused specifically on the healthcare enterprise and the achievement of healthcare affordability for all its primary facets: healthcare providers, medicine and pharmaceutical providers, healthcare machine and device providers, healthcare service and supply providers, and healthcare insurance and payment providers. The overarching goal is to accomplish a patient-centered environment for attaining healthcare affordability. The approach is achieved through motivation of people, improvement of processes, and increase of performance.

The four cornerstones for the foundation of healthcare affordability have been defined as a result of the work I've done with a multitude of the clients and industries I've served over the past 25 years. The first two of the cornerstones are leadership and strategy that resides at the strategic executive level of every organization. The third and fourth cornerstones for performance are lean and six sigma at the operational and tactical levels, using and utilizing their primary philosophies, methods and techniques for improvement, and most of all, they are the tools in two comprehensive tool boxes supplemented with the basic and advanced quality tools. Leadership and strategy sets direction, aligns the resources, and motivates the people. Lean and six sigma improves the processes and increases system performance.

Healthcare affordability is not theory, but a proven, fact-based approach, developed from the results of dozens of programs and projects, designed to meet customer requirements for value delivery, using the right resources at an optimized expense, and offered at a price that is competitive providing profitable revenue. What can be more beneficial than a faster, better, more affordable healthcare enterprise?

Paul Walter Odomirok, Sr.

chapter one

Making the case for Healthcare Affordability

This book is intended to answer a few simple questions:

- Why isn't healthcare becoming more affordable?
- How can healthcare become more affordable?
- What can be done to achieve Healthcare Affordability?
- Who needs to be involved to accomplish Healthcare Affordability?

Affordable healthcare and Healthcare Affordability

Affordable Healthcare continues to be an elusive goal as the expense of care continues to escalate and the investment itself yields a lower return. In addition, the third leading cause of death, following heart disease and cancer, are medical errors that increase cost, create waste, escalate risk and liability, and consume valuable resources that could be applied to quality care and patient care accessibility. Affordable Healthcare cannot be isolated nor consolidated to the multitude of singular and solitary subjects such as coverage, condition, outcomes, mandates, accessibility, and cost. At this point in time, it has proven to be true, Affordable Healthcare cannot be legislated in terms of how it is, how it has been, or how government wants it to be. Healthcare is about caring for, and healing patients.

Healthcare Affordability, on the other hand, can be achieved through the constant, relentless, and continuous pursuit for faster, better, more affordable, patient-centered care. In making the case for Healthcare Affordability, one must think differently. It's not about "in the box thinking," nor "out of the box thinking," it's about "new box thinking." Creating a new paradigm of thinking, strategizing, and operating using a completely new archetype. In fact, it goes beyond the boundaries of the current state of the domain of care, encompassing the entire Healthcare Enterprise that includes: Healthcare Providers, Medicine and Pharmaceutical Providers, Healthcare Machine and Device Providers, Healthcare Service and Supply Providers, and Insurance and Payment Providers. It is a systems approach, focusing on attacking the root cause of the many problems, and providing solutions for increasing care delivery speed, responsiveness, quality, capacity, accessibility, and affordability.

Below are several opinions, from various sources, on the current state of Healthcare:

- According to *Forbes* (Leah Binder), "The Five Biggest Problems in Healthcare Today"[1]; (1) Too much unnecessary care, (2) Avoidable harm to patients, (3) Billions of dollars are being wasted, (4) Perverse incentives on how we pay for care, and (5) Lack of transparency
- From HealthPAC online[2]: The issues to address in the reform of the healthcare system: affordability, portability, accessibility
- CNN's report on "America's 9 Biggest Health Issues"[3]: Doctor shortage, hospital errors and infections, antibiotic resistance, more do-it-yourself healthcare (apps and technology), food deserts, caregivers for the aging population, the cost of Alzheimer's, marijuana, and missing work–life balance
- *Healthcare Business and Technology* published "Top 10 Issues impacting the Healthcare Industry in 2016"[4]: mergers, drug prices, mobile care, cyber security concerns over medical devices, money management, behavioral health moves to front of stage, community care collaboration, new databases, welcome biosimilars, and medical cost mystery.
- Slate's Medical Examiner[5] tells us: Congress Has Forgotten America's Biggest Healthcare Problem—The critical question is not who gets care, and who doesn't, but how it's delivered.
- The Physicians for a National Health Program[6] claim: (1) Americans pay way, way, way more for health care than anyone else. (2) We pay doctors when they provide lots of care, not when they provide good health care. (3) Half of all health care goes toward 5% of the population. (4) Our health insurance system is the product of random WWII-era tax provisions. (5) Insurance companies have small profit margins. (6) Getting health care in the United States is dangerous (medical errors). (7) One third of health care spending isn't helping make Americans healthier (i.e., $765 Billion). (8) Obamacare is not universal health care.

Many of these opinions are quite a bit more qualitative than quantitative. However, a few are actually facts based. Regardless of the level of the use and utilization of data, it is apparent that there are many opinions, from many sources, for many reasons, with many "agendas." All pointing in the same general direction.

Whether the observations are by opinion or by fact, it is clear to many that we are heading in a direction that is not making conditions better for the population. We're spending more and more on a system that is returning less and less. When you consider how our health investment compares to our life longevity with the rest of the world, you find that we are spending more, and getting less. So let's take a look at some facts (Figure 1.1).

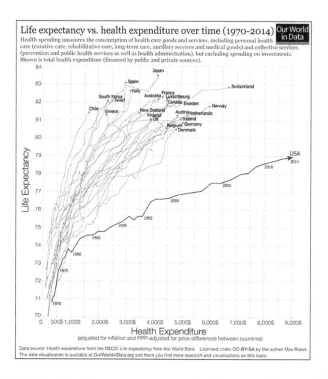

Figure 1.1 Worldwide life expectancy vs. health expenditures. (Data from https://www.ineteconomics.org/perspectives/blog/the-link-between-health-spending-and-life-expectancy-the-us-is-an-outlier.)

After this chart was created, our per capita spending in 2015 was $9,900, and our 2016 spending came in at $10,345, while life expectancy was barely affected. Obviously, our Healthcare has not gotten more Affordable, but instead, much more costly, with less of a positive effect for our future. In fact, when we compare our system with the rest of the world's economic leaders in terms of GDP, you'll find the gap is widening, while the quality of care is not. Once again, we are paying much more as a portion of our overall wealth, with a system that is not that much greater on a per person scale. If we go back in time, we'll find the per capita cost in healthcare in 1960 was $147.00, in 1970 $356.00, continuously growing until to $7,911.00 the Patient Protection Affordable Care Act (PPACA) was created in 2008 (Source: Henry J. Kaiser Family Foundation[7] with 2009—$8,149.00 and 2010—$8,402.00). Since that time, according to HHS and OECD, 2014 came in at $9,442.00, 2015 ended at $9,900.00, and 2016 capped the cost explosion at $10,345.00.

From yet another perspective (see the graph below), using the GDP, we can compare the amount of wealth that is spent on healthcare worldwide.

From 1980 to 2013, most of the developed countries around the world were spending between 5% and 9% of their GDP in Healthcare. Almost 35 years later, the U.S. is spending over 17% while the rest of the "pack" are spending between 9% and 12%. Clearly, the gap has widened (Figure 1.2).

Looking back to 1960 and 1970, we were spending 5.0% and 7.3% of the GDP on Healthcare, respectively (Source: Bureau of Economic Analysis[8]). The forecast is bleak. Soon we will be spending 20% of our GDP, and when extrapolated, following the same path we're on, 20% or more is possible. So, if we were getting our money's worth, it would be safe to assume, spending more dollars, investing more of our economic wealth, our care should be best in the world.

From a technological, capable and quality perspective, the U.S. should be the best. After all, we have the best capability for highest quality health care through the most advanced technology available, and the best trained nurses and doctors. Below is a snap shot was taken by the OECD in 2013 of Health Date taken of 11 different countries across the world who provide quality care (Figure 1.3).

The U.S. did not come in first or second in any of the categories. However, it did come in tenth and eleventh in many of the categories. Our system is not getting better. One approach would be to attack and heal the system in the areas that are easily under our control, such as medical errors. Medical errors are quite costly, in terms of care, in terms of risk and litigation, and even in terms of resources lost addressing errors that could be applied to providing quality care. In fact, medical errors are one of the leading causes of death in the U.S. (Figure 1.4).

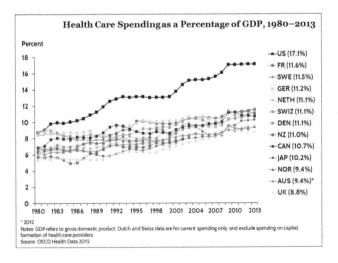

Figure 1.2 Healthcare expenditures as a percentage of GDP. (Data from OECD Health Statistics 2015.)

Worldwide Healthcare Rankings

COUNTRY RANKINGS
Top 2*
Middle
Bottom 2*

	AUS	CAN	FRA	GER	NETH	NZ	NOR	SWE	SWIZ	UK	US
OVERALL RANKING (2013)	4	10	9	5	5	7	7	3	2	1	11
Quality Care	2	9	8	7	5	4	11	10	3	1	5
Effective Care	4	7	9	6	5	2	11	10	8	1	3
Safe Care	3	10	2	6	7	9	11	5	4	1	7
Coordinated Care	4	8	9	10	5	2	7	11	3	1	6
Patient-Centered Care	5	8	10	7	3	6	11	9	2	1	4
Access	8	9	11	2	4	7	6	4	2	1	9
Cost-Related Problem	9	5	10	4	8	6	3	1	7	1	11
Timeliness of Care	6	11	10	4	2	7	8	9	1	3	5
Efficiency	4	10	8	9	7	3	4	2	6	1	11
Equity	5	9	7	4	8	10	6	1	2	2	11
Healthy Lives	4	8	1	7	5	9	6	2	3	10	11
Health Expenditures/Capita, 2011**	$3,800	$4,522	$4,118	$4,495	$5,099	$3,182	$5,669	$3,925	$5,643	$3,405	$8,508

Notes: * Includes ties. ** Expenditures shown in $US PPP (purchasing power parity); Australian $ data are from 2010.
Source: Calculated by The Commonwealth Fund based on 2011 International Health Policy Survey of Sicker Adults; 2012 International Health Policy Survey of Primary Care Physicians; 2013 International Health Policy Survey; Commonwealth Fund National Scorecard 2011; World Health Organization; and Organization for Economic Cooperation and Development, *OECD Health Data, 2013* (Paris: OECD, Nov. 2013).

Figure 1.3 Worldwide healthcare rankings. (Data from International Health Policy Survey of Sicker Adults (2011) and Primary Care Physicians (2012), Commonwealth Fund National Scorecard (2011), WHO, and OECD Health Data (2013).)

Some personal observations and case-study perspectives

I've been serving the healthcare industry for 20 years. When I get engaged with a Healthcare organization, there are numerous wastes that are inherent in our healthcare systems. Specifically, there are three that typically appear:

- Waiting: Patients waiting, patient care givers waiting, healthcare providers (i.e., physicians, nurses, technicians, assistants, and aides) waiting, machines and devices waiting to be utilized, with materials and supplies waiting to be used.
- Excessive motion: Systems designed to cause workers to expend energy in movement and motion, in order to, "get the job done." For example, nurses having to spend a great deal of their time going to storage locations to get materials and supplies to treat patients.
- Medical defects: Being the third leading cause of death, flaws, and weaknesses in systems as a result of inconsistencies and imperfections of processes and procedures.

When addressing improvement opportunities, numerous institutions often follow similar traditional, conventional approaches. Many of the

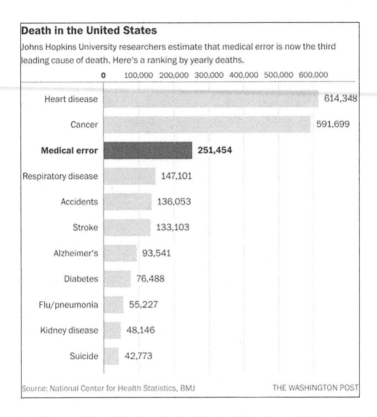

Death in the United States

Johns Hopkins University researchers estimate that medical error is now the third leading cause of death. Here's a ranking by yearly deaths.

	Deaths
Heart disease	614,348
Cancer	591,699
Medical error	251,454
Respiratory disease	147,101
Accidents	136,053
Stroke	133,103
Alzheimer's	93,541
Diabetes	76,488
Flu/pneumonia	55,227
Kidney disease	48,146
Suicide	42,773

Source: National Center for Health Statistics, BMJ   THE WASHINGTON POST

Figure 1.4 Cause of death in the United States. (Data from National Center for Health Statistics, BMJ.)

following symptomatic conditions are at play when it comes to performance challenges, waste elimination, defect and variation reduction, process improvement, and continuous improvement:

- Lots of meetings, with lots of people.
- Lots of charts and spreadsheets, with lots of data.
- Lots of emails, copying lots of people.
- An inherent "shame and blame" culture, trying to find out "who did it."
- "Busy-ness" with lots of effort and lots of activity.
- A cautious, defensive behavior by those directly involved.
- A Physician-centered System based primarily on Physician Decisions and Physician Schedules.
- A Revenue Cycle Focus putting money as the top priority for the organization.
- …and lots more.

While conditions for improvement efforts have gotten better over the past 20 years, much of the same conditions often linger in pockets and places throughout the culture. During the past 15 years, approaches have advanced in quite a few areas:

- Assign and task individual with improvement activities (i.e., more accountability for "fixing").
- Increased awareness of focus on quality.
- More performance feedback has been emerging.
- New policies and procedures focused on "fixing the problem."
- Some incentives for resolving quality issues.
- Punishment and penalties continued to prevail.
- The emphasis to promote good will emerged.
- The attitude to work harder to do better was embraced.
- Four common themes for poor performance (Roger K. Resar, Making Noncatastrophic Health Care Processes Reliable: Learning to Walk before Running in Creating High-Reliability Organizations[9])
  - Emphasis on hard work and vigilance
  - Benchmark of 50% set a false sense of process reliability
  - Permissive attitude toward variation
  - Processes not designed to meet outcomes

New approaches have been emerging over the past 5 years thanks to the efforts of several professional organizations including; I.H.I. (Institute for Healthcare Improvement), A.M.A. (American Medical Association), C.M.S. (Centers for Medicare and Medicaid Services), A.H.Q.A. (American Health Quality Association), and more. Over the past 5 years, I've been personally involved delivering professional development by way of a program with the IISE (Institute of Industrial and Systems Engineers)[10], sponsored by the Georgia Hospitals Association for Hospitals in the State of Georgia. It is an approach using use Lean and Six Sigma tools, techniques, and methods to increase the performance of the nearly 200 hospitals across the state of Georgia. Several hundred healthcare professionals have been involved, and the results of a few dozen projects have accumulated more than $40 Million in cost reduction, with increased speed and responsiveness in care, and substantial improvement in quality.

Nearly everyone I encounter, have memorable stories of Healthcare OFI's (Opportunities For Improvement). Everything from minor medical mistakes, to major medical errors, affecting lives. I've compiled ten case examples of my own, from first-hand experience, that elucidates the need for change.

- Case 1—Cost: Even as I write this book, my personal healthcare Insurance premium has increased, although I've never had a major claim, nor met the high deductible of the policy I've chosen. Recently,

I received a letter from my healthcare insurance provider, dated July 26, 2017, notifying me of an annual healthcare insurance increase of over $4,000.00. Part of the letter states,

We are writing to let you know that your rate is going to increase ... (a portion of the) increase in premium is attributable to the Federal Patient Protection and Affordable Care Act (PPACA) ... (the letter ends with the statement) Our commitment to you remains the same—to provide high quality health coverage and superior customer service.

Nothing in the letter is stated about their commitment to an effort focused on lowering healthcare cost or premium prices, nor working to improve healthcare in the U.S., let alone improving the speed and quality of care. However, today February 14, 2018, my Healthcare Insurance Provider, Golden Rule, a United Healthcare Company, is now charging me more for my healthcare insurance on a daily and yearly basis than the amount I pay for my food at the grocery store. In fact, my healthcare insurance payment is more than my mortgage payment. According to cost factors alone, healthcare is my greatest cost, and my number one investment ... and I'm not even close to my end of life. With my large deductible, I pay for most of my care out-of-pocket, and my insurance company pays in very little. To me, this seems a bit like extortion.

- Case 2—Hospital acquired infection: About 25 years ago, when my children were young, we had a snowstorm in the Atlanta area, and we decided to go down to a hill in the neighborhood for sledding. After several runs, I discovered a small boulder under the snow that opened up a wound on my left shin. It was deep (to the bone) and it required several stitches to close the tear in the flesh at the local emergency room. Since I had two layers of clothing, and my leg was not exposed to elements, the wound appeared clean, but in need of repair. Several days, and even weeks later, the wound was not healing properly. After going to the doctor, and receiving treatment and antibiotics, it was apparent, the wound was not cleaned properly. Today, I carry a scar, a reminder, of the poor quality of care.
- Case 3—Colonoscopy and endoscopy: "We can't make money doing both at the same time." I have a relative who requires a colonoscopy and endoscopy every several years. At one time, both were done simultaneously, for both convenience and cost reasons. Most recently, the Gastroenterologist refused to do both, stating that the system does not permit him to make money unless the two were separately performed. I changed to another provider.
- Case 4—Medical diagnosis errors and sepsis: "You're the one who called off the surgery." While working with numerous healthcare professionals at the Georgia Hospitals Association Headquarters,

on Monday, September 18, 2012, I received an emergency phone call from my brother in Florida. My father was admitted to the ICU in Port St Lucie Florida, suffering from severe back and abdominal pain. Just the night before, he and I were on the phone discussing the football games of the day, and at the end of our conversation, he told me he was having pain, and would call my brother to take him to the doctor. My brother brought him to the emergency room, and as a result of a doctor's diagnosis, he was scheduled for back surgery on Thursday. As I discussed his case and condition with the physicians I was working with, they gave me a list of questions to ask the hospital regarding my father. At the end of the week, I rushed to my father's side, and upon arrival, the ICU nurse met me in his room stating, "So, you're the one that called off the surgery." Of course I hadn't, but it was apparent, the hospital was in a shame and blame mode of care. To make a long story short, 2 weeks later, he was taken to emergency surgery with a ruptured colon, with an ultimate outcome of sepsis. He fought hard for a year, but lost the battle in September 23, 2013. Medical errors took his life.

- Case 5—Aortic Stenosis: "We need to do this now." I have a female relative with a diagnosed condition of Aortic Stenosis. This was identified several years ago. Upon discovery, the Cardiologist and Thoracic Surgeon insisted on an immediate valve replacement. A second opinion was sought, which concluded a different diagnosis; continuously monitor over time and address replacement when it is necessary. After years and years of reviewing and monitoring conditions, there has been no change and no necessity to have the heart valve replaced. Inconsistent diagnosis, and procedures delivered too early, continue to create waste, medical defects and escalated cost.

- Case 6—Three colonoscopy's in two days: Another relative, at the age of 90, was admitted to a hospital with lower abdominal pain. Given that she had fought and won over colon cancer 10 years earlier, it was decided that she should receive a colonoscopy to be sure that a similar condition was not occurring in another location. Since the first try was not clear, a second was ordered (turning out "inconclusive"), with a third ordered for the next day. Even a novice would question, why would you put a 90-year-old woman through three colonoscopy's in two days? Although, I admit, I am not a medical professional, any novice would question why you would put a 90-year-old woman through three colonoscopies in less than 48 hours. Physical fatigue, dehydration, potential leakage and sepsis, and mental anguish are all potential failure modes of such a practice.

- Case 7—Pneumonia and readmissions: A year and a half later, the 90-year-old woman in Case 6 had turned 92. On December 26, 2016 she was brought to the emergency room (Note: Her Doctor's Office was closed), suffering from severe fatigue, weakness, and flu-like symptoms. They examined her, prescribed medicine, and sent her home. A couple of days later, her conditions worsened. She was brought to the same location, the same thing happened, they examined her, prescribed new medicine, and sent her home. On New Year's Eve, believe it or not, her conditions worsened. The hospital admitted her, with a diagnosis of pneumonia, and she spent the next 20 days (most of January 2017) in the hospital and at the rehabilitation center. Since she had not been admitted, the re-admission penalty did not apply. The Healthcare Provider dodged a bullet, and reaped the financial benefit from Medicare for the 20 day stay in the hospital.

- Case 8—A "No Charge" procedure charged and billed: Last year, during my annual checkup with my Dermatologist, he noticed a small dry patch lesion on my left shoulder and treated it with liquid nitrogen. He told me at the time, "No Charge." A month later I received a bill for the treatment, and spent the next month resolving the issue with his office. Finally, months later, my reimbursement was delivered.
- Case 9—"Your appointment has been cancelled": Just this year, before my annual visit to my Cardiologist, I called the office to verify the day and time. The scheduler told me, "That appointment has been cancelled." I asked, "When did you plan to tell me?" (since this was only a week before my visit. I had to reschedule my annual visit for 2 months later, "When he was available." Recently, I found out that the entire practice of eight Cardiologists, split from the local hospital that owned the practice, and created an independent practice due to several reasons (One being control and manipulation of schedules that did not focus on, and benefit patients).
- Case 10—Root Canal … Legislators in the way, requiring a specialist: Lastly, a case that exemplifies the interference and regulation of government negatively impacting care. My dentist informed me, that the State of Georgia, one of the few remaining states to do so, requires a separate, unrelated dental service specialist to perform root canals on dental patients. Even when, the dentist is qualified and experienced in performing root canal procedures.

The 10 cases mentioned above are but a small sample of the millions of defective cases that take place annually. The Conclusion: Time Wasted, Money Wasted, People Wasted, Care Services Wasted, Efforts Wasted.

Some positive considerations and conditions

The news is not all bad. From my experience around the world with leadership, strategy, and improvement projects, I have been able to experience healthcare around the globe. The one I have seen that provides the best care outside the U.S. is Japan. Although the Japanese system provides excellent care, the U.S.'s system has the capability to provide even better care than Japan. Accessibility, care quality and cost tend to be the areas where we fall short. This century, there has been a lot of attention paid in healthcare to defining quality, measuring performance, and more so most recently, improving processes and systems. Here's an example of Quality and Performance definitions and measurement systems focused on better healthcare;

- IOM Definition of Healthcare Quality:
  - Safe: avoiding injuries to patients from the care that is intended to help them.
  - Effective: providing services based on scientific knowledge to all who could benefit, and refraining from providing services to those not likely to benefit.
  - Patient-centered: providing care that is respectful of and responsive to individual patient preferences, needs, and values, and ensuring that patient values guide all clinical decisions.
  - Timely: reducing waits and sometimes harmful delays for both those who receive and those who give care.
  - Efficient: avoiding waste, including waste of equipment, supplies, ideas, and energy.
  - Equitable: providing care that does not vary in quality because of personal characteristics such as gender, ethnicity, geographic location, and socioeconomic status.
- Healthcare Quality Metrics and Measures (as defined by reputable Healthcare Associations, Commissions and Organizations):
  - The Joint Commission (TJC)
  - Center for Medicare and Medicaid Services (CMS)
  - National Committee for Quality Assurance (NCQA)
  - National Quality Forum (NQF)
  - American Medical Association (AMA)
  - Agency for Healthcare Research and Quality (AHRQ)
  - Utilization Review Accreditation Commission (URAC)

Since the 1990s, I've worked with many Healthcare Organizations, as well as providers of Healthcare Machines, Devices, Supplies, Services, Medicines, Pharmaceuticals, and Insurance. I've witnessed approaches and methods that have a positive impact, and now there seems to be the beginning of a positive paradigm shift in Healthcare that's occurring despite government and political interference and intervention. It has to do with leadership, strategy, people, solving problems, process improvement and increased performance. All attributes of Affordability.

Healthcare Affordability

Would it be reasonable to envision a future for U.S. Healthcare where Patient Care is timely, with higher quality outcomes, and much more cost affordability? A profound Healthcare Professional once told me, "Every person is a Patient. From the time they're born, until the time they die." I must impress upon every reader, the true picture is not all doomsday and dire. The possibilities, potential, and prospective for the future is bright. If, across the Healthcare Enterprise, we focus on the patient, provide complete service and support, increase the speed and accessibility, improve the quality and safety, and reduce the cost and price, there is hope for affordable healthcare. But it will take ALL elements across the entire Healthcare Enterprise.

I'll close this chapter with the snapshot below (see Figure 1.5) that clearly illustrates a system that is not sustainable, that does not have responsiveness, that is questionable in quality, and is obviously fraught with financial waste. It encourages me to ask;

- Do we have a system of Healthcare Affordability?
- How many resources, and how much time, is being Wasted?
- How much value, and how much quality care, is being Wasted?
- How much of the money spent is a Waste of Value and the Delivery of Quality Care?
- Why is Healthcare so penny wise and pound foolish?
- What do we need to do to achieve Healthcare Affordability?

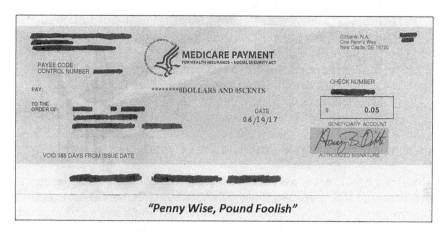

Figure 1.5 Very careful in small matters, very wasteful in large affairs.

References

1. Leah Binder, "The Five Biggest Problems in Healthcare Today", *Forbes*, 2016, https://www.forbes.com/sites/leahbinder/2013/02/21/the-five-biggest-problems-in-health-care-today/#75dc57a34587.
2. HealthPAC online, 2018, http://www.healthpaconline.net/health-care-issues.htm.
3. CNN's report, "America's 9 Biggest Health Issues", 2015, https://www.cnn.com/2015/01/02/opinion/gupta-health-challenges-2015/index.html.
4. *Healthcare Business and Technology*, Top 10 Issues impacting the Healthcare Industry, 2016, http://www.healthcarebusinesstech.com/issues-impacting-hospitals-2016/.
5. Nisarg A. Patel, Congress Has Forgotten America's Biggest Health Care Problem, Slate, 2017, http://www.slate.com/articles/health_and_science/medical_examiner/2017/07/delivery_not_access_is_the_biggest_problem_in_health_care.html.
6. Sarah Kliff, Vox, Facts That Explain What's Wrong with American Health Care, Vox, 2015, https://www.vox.com/2014/9/2/6089693/health-care-facts-whats-wrong-american-insurance.
7. Henry J. Kaiser Family Foundation, 2009, https://www.kff.org/health-costs/issue-brief/summary-of-coverage-provisions-in-the-patient/.
8. Bureau of Economic Analysis, 2015, https://www.cms.gov/Research-Statistics-Data-and-Systems/Statistics-Trends-and Reports/NationalHealth ExpendData/Downloads/HistoricalNHEPaper.pdf.
9. Roger K. Resar, Making Noncatastrophic Health Care Processes Reliable: Learning to Walk before Running in Creating High-Reliability Organizations, 2006, https://www.ncbi.nlm.nih.gov/pmc/articles/PMC1955343/.
10. IISE (Institute of Industrial and Systems Engineers), 2018, http://www.iise.org/.

chapter two

Healthcare Affordability defined

As time marches on, Healthcare Affordability becomes more and more and more important to our lives.

Healthcare Affordability occurs when care meets and exceeds the patient's requirements, delivered in a timely and high-quality fashion, at an affordable price, generating revenue that funds the application of the right resources at an expense that yields profitability. This encompasses not only the care provider, but also the cooperation, collaboration, and participation of those who provide healthcare support products and services; machine and device providers; service and supply providers; medicine and pharmaceutical providers; and finally insurance and payment providers. All five facets of the Healthcare Enterprise must be engaged, involved, and dedicated to serve the purpose (the Patient) and improve the processes and systems of delivering care, providing the machines and devices, providing the services and supplies, providing the medicines and pharmaceuticals, and providing the insurance and payments.

Improving processes and systems, involves solving problems that removes the root cause issue, increasing performance, and reducing variation. Some organizations believe that alleviating symptoms alone solves problems. Others believe, to truly solve a problem, one must eliminate the root cause of the problem. Unlike those who like to respond to symptoms alone, and there are many who chose to do so, I recommend focusing on the root cause and process performance, using a standard problem-solving process, for making decisions based on facts, balanced with the human factor of a team, discovering a resolution and solution that increases velocity, improves quality, and reduces cost. However, it is true, that using the problem-solving approach focused only on fixing symptoms demonstrates action, activity, and "busy-ness." But, by only patching symptoms, the waste of time, money, and resources lingers on. Solving problems using a problem-solving process, saves time, money, and resources. For example, in healthcare, treating pain symptoms reduces pain, but does it solve the person's medical problem? Pain is the result of a problem or condition that requires treatment or repair. Treating pain, and pain alone, does not fix "the problem," unless of course the problem is only the pain and pain alone. Resolving the problem should be the goal, not patching the symptoms. Healthcare Affordability is about solving problems for faster, better, more affordable Healthcare.

Healthcare Affordability basics

Healthcare Affordability improves the speed, quality, and cost of Healthcare delivered to patients (Suggested Reading; *Affordability: Integrating Value Customer and Cost for Continuous Improvement*, Chapter 1 "Affordability: It's Not What We Always Thought It Was!"). The factor of speed and velocity focuses on care delivery exactly when the patient needs it. Quality is about care that meets and exceeds requirements, expectations, needs, wishes, and wants of the patient. Finally, Healthcare Affordability cost should be, as Henry Ford might have put it, at a price that everyone can afford the care needed.

The speed of care must be increased to increase availability and accessibility. The quality of care must be improved to better patient outcomes. The cost of care must be decreased to decrease expense and reduce price. Each of these "musts" should be embraced by all dimensions of the Healthcare Enterprise. Care Providers, Medicine and Pharmaceutical Providers, Machine and Device Providers, Service and Supply Providers, and, Insurance and Payment Providers should all be guided, aligned, and motivated to adopt and institute the philosophy and practice of Healthcare Affordability.

Healthcare Affordability is not a static state. It's an evolving and ever-improving condition. More of a "journey" than a "destination." The focus on Faster, Better, and more Affordable Care across the Enterprise consolidates the overarching direction and goals toward Affordability. Think of it this way; You know you need care. (1) You get it now. (2) The care you receive solves the condition. (3) The "price paid" is reasonable and for the value delivered (Note: The term "reasonable," is the reality of perception of the patient.). Also, think of it this way; You have an unhealthy lifestyle. (1) You have the ability, and immediate accessibility, to the way(s) to improve your health. (2) The improvement methods are effective and beneficial. (3) The "price required" is worth the effort and energy for "value and quality of life."

Healthcare Affordability

In the core material on Affordability (see Chapter 1 in the foundational book on this subject, *Affordability; Integrating Value, Customer, and Cost for Continuous Improvement*), the 5P model was presented to describe the cyclical relationship of; Purpose, People, Process, Performance, and Profitability (see the Healthcare Affordability version of that model in Figure 2.1). As applied to Healthcare, the Purpose is the Patient, the People are those who directly provide the care and those who support the people who provide the care, the Process(es) refers to how the care and value is delivered, the Performance is based on the outcomes and results for the

The 5 P's as Applied to Healthcare Affordability
A Cyclical System and Function of Success

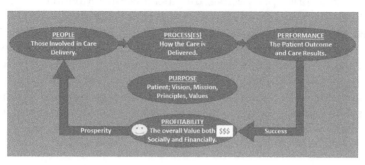

Figure 2.1 The five P's.

organization providing the care (including the outcomes and results of the business of the organization), and the Profitability from a financial and social purview. Each major component of this function is critical to Healthcare Affordability (Figure 2.1).

Purpose

The primary purpose is health and quality of life for everyone and every patient. Although this is a visionary perspective, it does provide an aim for the enterprise. In a traditional sense, the Patient is the aim. In an advanced sense, every person is the aim, and everyone, sooner or later is a patient of some type. Today, we often use, and state the term, patient centered care. However, we tend to continue to operate with an organization centered care approach, scheduling the resources (and even sometimes when conditions are emergent) under the convenience of time and availability (Examples include; Selective procedures done on the weekends for patient convenience (i.e., Dermatology Biopsies, Gastroenterologist Colonoscopies and Endoscopies, Orthopedic Examinations of Structural Minor Injuries, etc.). Although a 24 × 7 expectation may not be in the near future, it might be a visionary goal (Note: Although Hospitals operate in terms of 24 × 7, not all care services are generally available at all times, during all days of the week.). So, the core purpose of Healthcare Affordability is the Patient, and the Patient's "Point-of-Care" (POC; care at the place of the patient). One can imagine a day when POC is at the location of the patient, and not at the location of the institution providing the care. It may not be such a far-off reality with the advancement of technology and virtual reality.

People

From an enterprise perspective, this includes all individuals delivering the value of providing care, healthcare products and healthcare services, as well as those individuals supporting the delivery of care, healthcare products and healthcare services. People are the key drivers of the processes and systems within the entirety of the Healthcare Enterprise. I've often been told, Healthcare Affordability is very People-Centric, from a Customer Perspective ("Patient" and Care Givers), and a Provider Perspective (Those who deliver care and those who support care delivery personnel).

Process

"All work is a Process." The work of delivering care requires systems and processes specific to treating illness or sustaining health. It requires organizations and institutions focused on care, supplied by providers of machines and devices, services and suppliers, medicines and pharmaceuticals, and insurance and payment sources.

Performance

Although the primary performance focus is patient outcomes and results, underlying are the performance metrics and measures of the organization, the processes themselves and the people. Quantitatively, we can measure time, quality, and cost. Qualitatively, we can measure the degree of satisfaction of the patient, the people driving the processes, and all the suppliers involved directly and indirectly with the delivery of care.

Profitability

The financial dimension is always the emphasis, and on the forefront of the attention of providers. However, perhaps even more important, is the social dimension. In the Healthcare Enterprise, social profitability is palpable. When patient outcomes and results are stellar, motivation and people-performance is high. The inverse or reverse is also true: dismal results drive de-motivation and poor people-performance. I've been able to witness, first-hand, care organizations with positive performance and care organizations with negative performance. High-performing organizations are filled with motivated people. It should be intuitively obvious, poor-performing organizations de-motivate people, driving ineffective, and inefficient processes, with terrible patient outcomes.

Healthcare Affordability is about Faster, Better, and more Affordable Healthcare.

- Speed of Products and/or Services is/are achieved by
  - Delivering the Product or Service on time. Exactly when it's needed. Not too late. Not too early. Not too "much."
  - Delivering the Product or Service accurately. Exactly what is needed. Conforming to the Requirements. Complying with Standards. In a culture of continuous improvement.
  - Delivering the Product or Service at the location required. Exactly where it is needed. At the POC. In most cases, except when specific machines and tools are required, or a specific location is mandatory, the care should be applied where the patient is located.

Speed of Care achieved by

- Delivering Timely Care. When its needed. The "when" is the "now" in terms of the patient's need. That is, at the time and POC when its needed. This includes all the flow elements: providers (physicians, nurses, techs., etc.), medicines and pharmaceuticals, machines and devices, services and supplies, information, processes and procedures.
- Delivering Accurate Care. What is needed. The right treatment, at the right amount, for the right amount of time, exactly what is required.
- Delivering Accessible Care. Where it's needed. At the patient POC. With the minimal amount of; transportation, movement, motion, time, resources, medications, machines, etc.

Quality of Products, Services and Care achieved by

- *Conformance* with Patient and Customer Requirements, Needs, wishes, Wants, and Desires. Of course, patients are the care provider's customer, while the care providers are the customers of those elements supporting care (Providers of machines and devices, services and supplies, medicines and pharmaceuticals, insurance and payment). Clearly, in healthcare, the patient can assist with the symptoms and conditions (of course), but there is that "unknown factor" aiding in diagnosis and evaluation that needs to be supplemented by the healthcare professionals (i.e., doctors, nurses, specialists, technologists, all the "ologists"), machines and devices, services and supplies, medicines and pharmaceuticals.
- *Compliance* with Healthcare and Industry Standards: Regulatory, Industry Specific, Defacto, Designed, and Accepted by the Organization. This applies to the care providers, as well as all the other enterprise elements.

- *Continuous improvement* is the continual pursuit of Patient Centered, defect free, care with excellent outcomes. In addition, it applies to the continual pursuit for improving machines, devices, services, supplies, medicines, pharmaceuticals, insurance, and payment methods. Affordability prescribes

More Affordable Care is achieved by

- Delivery of Value that meets and exceeds Customer Requirements, Needs, Wants, Wishes, and Desires.
- Providing the right Resources at the right time to deliver Customer Value at a profitable Expense and Cost.
- Providing competitive and affordable Pricing to establish a profitable revenue stream that's affordable for Customers and beneficial to Providers.

Healthcare Affordability is the Delivery Value, per the Patient Requirements, at an Expense that provides the Right Resources, and at a competitive Price that generates profitable Revenue, balancing the triumphant Value, Patient, and Cost. As defined by Affordability, Value is in the purpose and delivery of what the customer requires, needs, wishes, and wants. Value can be extended to the entire community being served. Customer, in this case is the Patient, the patient's care givers, and the outcomes of the care delivered. Cost come from the two factors of expense for resources and service price for revenue and income. The Aim comes from Value, Patient, and Cost (Figure 2.2).

The term "Healthcare," typically refers to those directly engaged in providing patient care. As I've already briefly discussed, I have an alternative view from the enterprise perspective. Healthcare should refer to five elements: (1) those who Provide Healthcare Services, (2) those

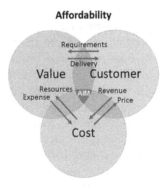

Figure 2.2 Affordability basics.

who Provide Machines and Devices used in Healthcare, (3) those who Provide Services and Supplies used in Healthcare, (4) those who Provide Healthcare Pharmaceuticals, and (5) those who Provide Healthcare Insurance and Payment. All five are inextricably linked to patient care. Healthcare Affordability is Patient Centered. The Affordability Aim of Healthcare Affordability is characterized by; (1) Delivering what the patients requires, needs and wants, (2) Using the right resources provided at a reasonable expense, (3) Offering products, services and care at a cost or price that is both equitable, and competitive, driving profitable revenue for keeping the organization in business and growing in response to increasing demand (Figure 2.3).

Healthcare Providers consist of organizations focused on delivering services to patients intended to improve health. Hospitals, clinics, private practices, and any organization designated by a state authority that provides healthcare. Their primary purpose should be to preserve and improve a person's health. This extends beyond just the typical medical services to dentistry, physical therapy, acupuncture, acupressure, chiropractic, and others, even physical training.

Healthcare Machine and Device Providers offer apparatuses, technologies, and equipment for Healthcare Providers to use in diagnosing illness and treating maladies.

Healthcare Service and Supply Providers offer services and supplies to support Healthcare Provider operations.

Healthcare Medicine and Pharmaceutical Providers offer drugs, medications, treatments, narcotics, analgesics, antibiotics, and other forms of remedies for Healthcare Providers to prescribe to Patients.

Figure 2.3 The healthcare enterprise.

Healthcare Insurance and Payment Providers provide the monetary flow to pay for the Healthcare Provider's services and supply their revenue cycle with funds. This type of provider comes in several forms:

- Private Policies that are paid for by the Patient, or more likely, a Patient's employer
- Public Policies provided by the Government (e.g., Medicare, Medicaid) or other funding institution funding public health (e.g., Shriner's Hospital, Free Clinics for the indigent)
- Supported Payments provided through others (Churches, Non-Profits, Charities, etc.)
- Self-Payment as paid for by the patient or the patient's family or the patient's care givers.

Using the Theory of Affordability, Healthcare Affordability utilizes the model of Affordability as the patient centered aim to influence and drive direction and alignment of the five Healthcare Enterprise elements. When all five elements focus on Healthcare Affordability, patient requirements are met, delivery performance increases, the right resources are provided at the right time, expense decreases, and the price can be reduced, influencing increased competitiveness, and profitable revenue flow, that maintains and sustains the organization success while growth and expansion occurs due to increased demand. Correct care, timely care, quality care, and affordable care are all attributes of Healthcare Affordability (Figure 2.4).

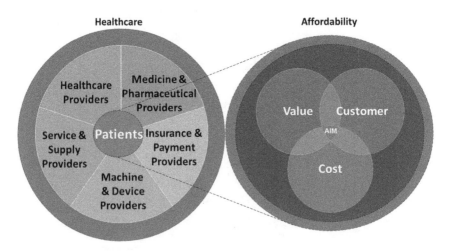

Figure 2.4 Healthcare and Affordability.

For achieving Healthcare Affordability, there are three main dynamics for all enterprise elements; Motivate People, Improve Processes, Increase Performance. As part of the 5P Model (Figure 2.1), People drive the Processes, and the outcome of the Processes results in Performance. This is not only true for the care providers but also for all providers in the enterprise (Figure 2.5).

Healthcare Affordability has a Patient focused core. All elements in the enterprise have the ultimate focus of patient. Patient is the primary focus and aim for the care providers. The other elements of the enterprise focus on delivering their products or services to the care provider that ultimately serves the patient. The care provider is in essence, a patient care integrator. This linkage assures and insures Affordability from the product and service providers, all the way downstream to the care providers, eventually delivering care to the patient.

The aim, similar to the IHI's Triple Aim, is about improving the patient experience of care (including quality and satisfaction), improving the health of populations, and reducing the per capita cost of health care. The difference is the scope and perspective. For a true and substantial improvement in all areas, the enterprise view of Healthcare Affordability lends itself to impact all upstream and downstream elements of the healthcare value stream (Figure 2.6).

The Healthcare Affordability Solution requires all elements of the Healthcare Enterprise to use all resources available to focus on providing machines, services, and care Faster, Better, and more Affordable. It is beyond change, its transformation. To achieve Healthcare Affordability,

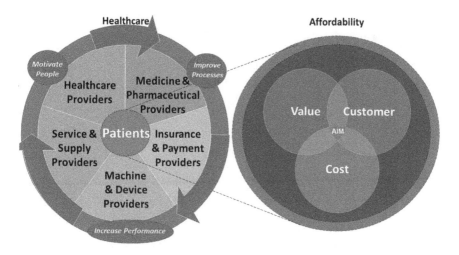

Figure 2.5 The path to Healthcare Affordability.

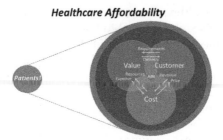

Figure 2.6 Healthcare Affordability is patient centered.

across the entire Healthcare Enterprise, in the words of Dr. W. Edwards Deming's 14 Points[1], "Point 14: Put everybody in the company to work accomplishing the transformation. The transformation is everybody's job." Motivate People → Improve Processes → Increase Performance (see Appendix A).

Reference

1. The W. Edwards Deming Institute, W. Edwards Deming's 14 Points, 2018, https://deming.org/explore/fourteen-points.

chapter three

Healthcare Affordability applied

> I do know this ... If you change the system, but the people aren't developed to take advantage of it, then nothing happens ... and, if you train people for change, but the systems doesn't allow it, then nothing happens either.
>
> **Paul Walter Odomirok**

The application of Healthcare Affordability requires that the system be changed from what it is today. Most definitely, it demands that the people be developed in conjunction with the system changes. The systems, being comprised of sets of linked processes, are driven by the people, and the performance outcomes and results are all in aggregation with the 5P cyclical system and function for success. Ultimate success is measured beyond the social and monetary dimensions of profitability. Profitability is only a result and outcome of successful performance. The measurement of success is based on three qualitative measures (customer satisfaction, people motivation, supplier satisfaction), three quantitative measures (time and speed, quality and reliability, cost and expense), and four overarching system scorecard measures (customers, business, process, people).

The qualitative measures are indicators of how the people involved view how an organization's value delivery is performing. The quantitative measures indicate how well the processes and systems are operating. And finally, the overarching system scorecard measures demonstrate an organization's level of achievement and accomplishment in terms of value delivery and purpose. Healthcare Affordability is realized when organizations motivate people, improve processes, and increase performance.

Traditionally, Healthcare has been defined around organizations and institutions that provide and deliver care for patients and clients of doctors, hospitals, clinics, and treatment centers, that is, locations of healing and curing any human illness. This purview separates care providers from the other elements of the enterprise. This "healing enterprise" encompasses all of the "How To's" that determine How to heal people. Healthcare Affordability stresses the importance of including all elements, and focusing on providing a compelling reason for all providers to participate in continual improvement. Just as important as doctors, hospitals, clinic, and treatment centers are the other providers focused on patient care (see Appendix A);

- Medicine and Pharmaceutical Providers
- Machine and Device Providers
- Service and Supply Providers
- Insurance and Payment Providers

Taking this approach and takt, a concerted focus on improving process speed and quality, with the inclusion of increased profitability and decreased cost, a better level of care will result, and favorable conditions for lower prices will emerge.

Many people don't believe that Faster, Better, more Affordable Healthcare is possible. They frequently ask the question, "Faster and better will cost more, won't it?" Or, the sometimes retort, "Faster, better or cheaper. Pick two!" For too many years, the perception has been that faster and better care will cost more. Over the past several years, I have been involved in pursuing faster, better, more affordable conditions in healthcare.

Table 3.1 illustrates just a finite set of the results of Healthcare Enterprise related projects. The table only displays the results from a cost reduction and savings perspective. Each project in the table also achieved improvements creating faster and better processes. Although most of these projects are of care provider origin, there are many listed that apply to the other four elements of the Healthcare Enterprise. The speed and quality aspects have not been included on purpose in order to "speak to the language of management" (i.e., money) and demonstrate the financial outcomes of efforts placed upon increasing velocity and improving quality. The projects listed below occurred during the time span of 2011–2017, by IISE Lean and Six Sigma Black Belt Candidates. These projects represent improvement activities from all types of organizations in the Healthcare Enterprise, and demonstrate the ultimate potential when Healthcare Affordability is applied.

The average savings, on a per project basis, amounts to more than $500,000 (i.e., $526,821.64). Cumulatively, the total savings is nearly $40,000,000.00 with a sample of only 70 projects. This set of projects represents a small subset of the vast opportunity that exists for reducing cost. In addition, with the projects listed above, yielding a reduction in process time for each (either lead time, cycle time, procedure time, or a combination thereof), with increased capacity and capability, and freed up resources, the overall result provides greater responsiveness and accessibility of care and, better availability of higher quality care-related products and services.

Table 3.1 Results of Healthcare Enterprise Related Projects

Project description	Annual savings
Reduce no shows at dental clinic	$72,270.00
Reduce unnecessary technician rolls	$5,200,000.00
MRI order turnaround time reduction	$700,000.00
Reduce cycle time for human resource requisitions	$789,597.00
Inventory optimization	$300,037.00
Reducing patient clinic wait times	$27,800.00
Bed cleaning turnaround time reduction	$24,219.00
Stockroom transaction error reduction	$373,273.00
Hiring cycle time reduction	$805,728.00
Warehouse receiving process improvement	$84,688.00
Calibration system redesign	$56,516.00
Adhesive improvement	$13,959.00
Scrap reduction	$38,000.00
Technical support call frequency reduction	$416,664.00
Repair cycle time reduction	$110,000.00
Supplier productivity	$600,000.00
Non-conforming material process improvement	$170,177.00
Call routing enhancement	$208,705.00
Insurance marketing pilot	$1,000,000.00
Data center management	$370,000.00
Improved scheduling	$57,810.00
Emergency department throughput	$3,022,000.00
Improving observation standards	$1,700,000.00
Mammography reporting process cycle time reduction	$517,000.00
Inventory management	$433,555.00
Pallitive care	$456,304.00
Retain BioBurden	$126,000.00
Facilities management request handling	$217,017.00
Employee access time reductions	$505,000.00
Blood culture contamination rates	$755,199.00
Service quality improvement	$988,836.00
Implanted tissue traceability	$190,000.00
Patient throughput (LOS reduction)	$500,000.00
Bed placement times	$575,000.00
Sepsis reduction	$575,000.00
Quick meds	$282,400.00
Reducing blood contamination	$328,800.00
Reducing clinic turnover rates	$160,560.00

(*Continued*)

Table 3.1 (Continued) Results of Healthcare Enterprise Related Projects

Project description	Annual savings
Reducing unbilled services at vet clinic	$39,000.00
Reducing urinary tract infections	$152,100.00
Reducing sepsis infection	$2,328,032.00
Reducing falls in hospital	$1,069,562.00
SFH floor Pyxis optimization	$70,000.00
Revascularation (reduce LOS)	$1,294,091.00
Increasing case mix index	$95,967.00
Decreasing the no show rate	$62,212.00
Reducing denier variation	$334,000.00
Transforming care delivery in emergency services	$373,500.00
Reducing contract overpayments	$466,000.00
Security patch reduction	$102,449.00
Enrollment revenue increase	$250,000.00
Inventory reduction	$341,532.00
Manufacturing scrap reduction	$298,660.00
Reducing P card administrative costs	$64,860.00
Reject rate reduction	$46,000.00
Reducing patient non-chargeable supplies	$150,291.00
Outsource help desk	$629,308.00
Increasing press capacity	$250,000.00
Reducing ED LOS	$48,283.00
Reduce average X-Ray turnaround time in the emergency department	$106,000.00
Orthopedic surgery scheduling	$390,000.00
Reducing clinic leave without being seen visits	$1,263,000.00
Denial prevention	$1,327,896.00
Inventory reduction	$175,958.00
Safe hospital	$55,000.00
Process improvement program	$187,000.00
Improving patient access to beds	$1,800,000.00
Assembly time reduction	$250,000.00
Improving shift scheduling	$96,000.00
Laboratory turnaround time reduction	$8,700.00
Total savings	$36,877,515.00
Average savings	$526,821.64

Source: IISE (Institute of Industrial and Systems Engineers).

In addition to improved quality, the valuable resources freed up provide for better delivery of care, through the use of better products and services used in care delivery. Primarily, these projects were executed in Healthcare Provider establishments, with the other projects completed in machine and device production, service and supply provisioning, medicine and pharmaceutical manufacturing, and insurance and payment administration. If we, "do the math," according to the AHA[1], there are 5,564 Hospital Systems in the U.S. If, in each system, only one project was executed, $2,931, 235,604.96 could be saved … what about 2 projects? ~ $6 Billion … what about 4 projects? ~ $12 Billion. Now take that across all 5 Healthcare Enterprise elements ~ $60 Billion. Although this is only the beginning, scratching just the surface, hundreds of billions, if not trillions in savings can be harvested. Our Cost-of-Care would decrease, and our Affordability would dramatically increase.

So the obvious question confronts us. What can be done to influence the entire Healthcare Enterprise to focus on Healthcare Affordability?

- Stop rewarding organizations on the basis of money, and money alone. And, above all, do not bail out the failures, which is just a reward for bad practices, bad behavior, bad management, and bad leadership.
- Start providing incentives emphasizing the increase in process speed, the improvement of quality for products and services, and the reduction of cost and expense.
- Encourage the leaders of all affiliated organizations to set direction, align resources, motivate their people, communicate the message, and execute the plan to achieve Affordability.
- Strategically, operationally, and tactically, motivate people, improve processes, and increase performance.
- Finally, put rewards in place for conformance to requirements, penalize non-conformance, incentivize compliance to standards, correct non-compliance, and promote continuous improvement.

Creating and applying a Healthcare Affordability

Healthcare Affordability is neither a treatment nor a remedy. It's not some magic potion devised to cure corporate maladies, nor an antidote for company sickness. However, it does provide the framework and foundation for using existing resources to create solutions to solve problems that exist within any enterprise. As with any contagion, if contracted, it should spread through the "organism" (i.e., organization), and throughout the community (i.e., the Healthcare Enterprise). One of the barriers and roadblocks comes from resistance: resistance to change and transformation. Creating and applying Healthcare Affordability requires a purpose, a vision, a leader, a strategy, people, resources, processes, and a design and plan.

Healthcare Affordability Applied

Figure 3.1 The application of Healthcare Affordability.

There are three levels that exist in design and implementation: strategic, operational, tactical (see Figure 3.1). One is no more important than the other, and not one is least important of all. Failures of change and transformation frequently occur between leadership, management, and the people. Disconnects in alignment or fractures of communication often ruin beneficial initiatives and valuable strategies. Systems, processes, and value delivery regularly breaks down when all levels are not fully engaged and involved. The communication continuum from top to bottom, and bottom to top, is ineffective if there are fissures in the structure and systems. The pyramid, traditionally, is a sound and stable structure. However, if not tended to, cultivated, and continuously improved, it tends to crumble.

The upper most level, called "Strategic," includes the top leadership and senior management. The many roles of leadership includes: move the organization forward, maintain competitiveness, sustain relevance, foster growth, incorporate necessary change, cultivate focus, set direction, align the resources, motivate the people, communicate the message, and execute the plan. Critical and key performance areas for success are found in Kaplan and Norton's Balanced Scorecard: Customer, Business, Process, and People. When strategy is deployed, leadership must also institute a rhythm of performance review at all levels that checks status, performance against goals and objectives, efforts to solve problems, endeavors of improvements, and necessary resources requirements.

The middle level, often referred to as "middle management" or "Operational," is comprised of managers and employees who manage the processes of the organization. This involves the fiduciary functions, the intellectual property, the information technology, the physical assets, the resources, the value production capabilities (both in terms of products and services), the delivery methods, the supply chain, and any other function needed to support value production and delivery.

The base level, titled "Tactical," is where value creation and value delivery happens. It encompasses the staff of people that either provide value or support the creation and delivery of value. It is the zone of value work, the part of the organization that is doing the things the customers are willing to pay for and purchase. If put in the proper perspective of hierarchy of value, this level should be placed at the top, instead of on the bottom.

A key facet of this design is the inextricable linkage from top to bottom, and bottom to top, of purpose, direction, alignment, motivation, communication messaging, and execution of the plan. This linkage insures and assures that gaps, breaks, and interrupts in the structure and systems do not occur. Two tools that should be employed are regular organization performance reviews and "Catch Ball." Organization performance reviews, with regularity of rhythm, check status of process and people performance, performance to goal, the results of improvement efforts and resources required to accomplish the opportunities and challenges that exist. Catch Ball is a technique of passing information throughout an organization from top to bottom, and bottom to top in order to guarantee consistency and integrity of intelligence, knowledge, messaging, reporting, and policy. It involves communication and feedback down and up for the purpose of transparency and consistency.

A key challenge in this design is a condition of "Natural Tension" between Strategic Leadership and Operational Management. Leadership is about change and dynamics of the organization. Management is about status quo, stability, static, consistent, and unchanging functions and procedures. The purpose of each tends to create a natural tension between the two levels. If not addressed correctly, the tactical or people level will recognize this friction as conflict, and with apprehension, throttle their efforts. Although this is natural, if not openly addressed and mitigated, it could create negative results, thereby slowing progress.

The role of the People at the tactical level is to either deliver value to the customer, or provide support for those who deliver customer value. Leaders lead the organization at the strategic level. Managers manage the processes and system components at the operational level. The people deliver value to the customer at the tactical level. The roles and responsibilities are clear.

The Healthcare Affordability construct and context can be defined, but, before applying Healthcare Affordability to any organization, one must understand the current state, design a solution, implement the solution, and ultimately maintain the design.

Assess → Design → Implement → Maintain (for this section, please refer to Appendix B)

This is a four-step redundant pattern for attaining success and realizing momentum. Where each step ends, the next step begins. The "last step" (i.e., Maintain) is just the preceding step for the next assessment, and the next round of activities. It can be considered the Affordability Algorithm. It is made of four sequential steps or stages. The initial stage Assess is about researching and understanding the current conditions and identify the focus areas for improvement. The Design stage provides a blueprint and plan for action. The Implement phase is for Initiation and realiza tion of the design. Finally, the Maintain step stabilizes and standardizes the new paradigm, and serves to "front-end" the next assessment stage in order to replicate the pattern again.

Assess

> If you do not know where you come from, then you don't know where you are, and if you don't know where you are, then you don't know where you're going. And if you don't know where you're going, you're probably going wrong.

> **Terry Pratchett[2]**

Often, organizations jump to conclusions and begin implementing before spending time understanding, designing, and planning. This level of urgency typically returns a burden of more time and resources spent and wasted. The outcome is usually short of expectations and disappointing.

To increase the probability of a good outcome, research and investigation must take place to establish a good understanding of the current state of Purpose, Value, Customers, Costs, Direction, Leadership, People, Resources, and Tools. This knowledge serves to help shape the design and plan. It also assists in bringing the design team to the same level of awareness of where the organization exists today. Such intelligence provides the basis for shaping targets, goals, and objectives, and often discloses opportunities for improvement.

Design

> Form follows Function.

> **Louis Sullivan[3]**

Architecture of the 20th and 21st centuries was shaped in the 1800s by Louis Sullivan. Louis Sullivan was a mentor of Frank Lloyd Wright and known as one of "the recognized trinity of American architects." His message resounds when it comes to designing a Healthcare solution.

Traditional paradigms of Healthcare are fraught with designed-in system defects, ineffective flow constraints, and waste in terms of waiting, motion, and medical defects.

By designing solutions and improvements with the knowledge of patient and system requirements, more efficient and effective processes can be implemented. This is not just for physical spaces but also for procedural conditions such as material flow and supply accessibility. It is also for information access and treatment procedures. The design must proceed the plan and schedule for implementation.

A great deal of focus and attention must be paid to team planning and scheduling for implementation. This does not directly imply that a great deal of time is required. However, it does suggest that a comprehensive plan and schedule should be created, and adequate time should be applied. Once the design, implementation plan, and schedule are complete, the solution is ready for implementation.

Implement

> You've got to eat while you dream. You've got to deliver on short-range commitments, while you develop a long-range strategy and vision and implement it. The success of doing both. Walking and chewing gum if you will. Getting it done in the short-range, and delivering a long-range plan, and executing on that.
>
> **Jack Welch[4]**

There's power, energy, and motivation in "Short-Term Wins." There's integrity, strength, and influence from "Long Range Accomplishments." Both must occur for paradigm changes and transformations to take hold and be fulfilled.

Short-Term Wins establish a confidence in leadership and the people that they can attain and achieve success. Short-Term wins can be as simple as procedural changes for improving efficiency. Short-Term Wins can prove to teams that they can "do it." A portfolio of Short-Term Wins can create change and move forward to improve the organization.

Long-Range Accomplishments should follow Short-Term Wins. Long-Range Accomplishments should always be included in the design and plan for implementation. Long-Range Accomplishments serve to establish and institute new actions, behaviors, and standards during a transformation. The long term perspective establishes a strong sense of direction, alignment, and motivation in the people. It sends the message that leadership is committed to the future.

Combining the short term with the long term lays out a continuum of success and establishes confidence in the people. A stable, standard, sustained environment, with momentum and success, that shifts and adapts to new challenges and opportunities, establishes a setting in which people thrive. Maintaining such a condition is difficult and demanding. Leadership must maintain the gains while also designing and planning for the future.

Maintain

> Maintaining success, and sustaining momentum, is not the end of an effort, but the beginning of an emerging era.
>
> **Paul Walter Odomirok**

The state of "Maintain" is not static. It is dynamic in that it requires that maintenance and sustainment eventually return to the "Assess" step and cycle through "Design" and "Implement" again. To maintain, is to maintain the everlasting continual improvement effort of Assess → Design → Implement → Maintain. Similar to PDCA or PDSA for process improvement and problem solving, it's a cyclical function that is repetitive and redundant. A continuous improvement culture can be instituted by following these four simple steps, and repeating them forever.

The duration

I'm often asked, "How long does it take?" I will typically respond, "It could take 3 to 5 years or more, and sometimes 5 to 8 years for some organizations to complete stage one. The initial effort typically takes 9, 12, and often 18 months." The "full truth answer" is: Forever. Without exasperating those who ask, I try to influence their understanding of the duration of the initial step, without frustrating their enthusiasm with the extent of the ultimate reality. Affordability is an ongoing, continual evolution of change and transformation.

Affordability planning and initial deployment

I've prepared an initial implementation template, spanning 16 months of inaugural actions and activities, front ended by 2 to 3 months of planning and setup, and succeeded by the next steps and stages of change and transformation (see Appendix C). Some organizations take 6 to 9 months for their first phase of action. Some take 12, some 18, some more. The 18 month window is appropriate for realizing both short-term wins and long-term accomplishments.

A core team may spend as much as several months in an upfront assessment effort to obtain the vital information and data to begin. This core team would consist of 5 to 8 individuals with subject matter expertise in various areas for the purpose of creating a comprehensive portfolio of intelligence to serve as the basis for configuring, design, planning, and scheduling the initiative.

A guidance team for this effort should be established using top leadership advocates for this endeavor. The guidance team will provide direction, establish alignment of resources for the activities, motivate the participating individuals, serve as communicators of the message, and ensure execution of the plan. The core purpose of the guidance team is to provide guidance for the design team and the execution teams for the projects and activities.

The design team is comprised of influential individuals who demonstrate strong support for the effort and can get things done. Their role is to deploy design, per guidance from the guidance team, with responsibility and ownership of projects, activities, and actions. It is expected that design team members participate in design, planning, and scheduling. The purpose of the design team is to lead the people involved in the efforts as planned, and to participate and engage in the implementation. It may be plausible for each design team to expect to lead multiple implementation teams and efforts. This team provides the core expertise and support of the entire initiative.

Purpose, direction, vision, mission, and goals are initially established by leadership, and through the use and utilization of the "catch ball" process, honed and shaped to create a collaborative aim for the initiative. During the initial deployment, communication, emphasis, review, and adjustment for these alignment elements is ongoing. This incorporates the guidance team, the design team, and the participating people.

The people participating should be carefully selected from within the organization be the guidance and design team. Each person must be evaluated on ability, capability, commitment, enthusiasm, and attitude so as to avoid selecting resisters and inhibitors. These individuals should be trained and developed for their role or roles over the implementation horizon. Their involvement should be positioned in a manner that the activities of the initiatives that they are involved with is part of their job, not work assigned in addition to their job.

Processes and resources, and performance measurement and communication are the enabling factors for this initiative. It is the responsibility of the guidance team to see to it that all of these dynamics are in place. It is the responsibility of the design team to take advantage of, and utilize, these initiative components. The people have the permission and approval to access and use these aspects and features of the program for implementation.

Summary

Applying Healthcare Affordability requires a substantial commitment by the organization's leadership. All of the top leadership have to commitment themselves and sign up for full accountability for the result. It involves strategy, structure, and systems, but above all, it takes strong leadership. Although the effort mantra is simple and straightforward (i.e., "Motivate People → Improve Processes → Increase Performance), it takes time, energy, and commitment to accomplish the goal. If organizations treat it like just another "strategy du jour," failure will result. If organizations don't follow the discipline for institution, the effort will be a disaster. If organizations use this as an excuse for downsizing, the results will be dismal. However, for those who have done it correctly, success is eventual.

The key points for application are as follows:

- Required changes: philosophy and culture, purpose, strategy, systems, structure, people.
- Affordability alignment: from the customers through the corporation through the suppliers.
- Improvements: many quick wins, with numerous strategic, operational, and tactical initiatives.
- Focus: faster, better, more affordable, processes, products, and services.
- To improve (continually): assess → design → implement → maintain
- Duration: overall → eternity, initially → typically 12 to 18 months.

References

1. AHA, 2018, https://www.aha.org/statistics/fast-facts-us-hospitals.
2. Goodreads, Terry Pratchett Quotes, https://www.goodreads.com/quotes/412254-if-you-do-not-know-where-you-come-from-then.
3. *Encyclopedia Britannica*, 2017, https://www.britannica.com/biography/Louis-Sullivan.
4. Larry Kudlow, An Interview with Jack Welch, 2009, https://www.nationalreview.com/kudlows-money-politics/interview-jack-welch-larry-kudlow/.

chapter four

The strategic dimension

At the strategic level are leadership, strategy, performance ... guiding and steering the organization.

It all works the best when leadership leads the way, sets the strategy, and motivates the people, thereby setting the direction, aligning the resources, executing the plan, and increasing performance. It works best when management manages the systems and processes, focusing on performance improvement. It works best when the people deliver the value to the customer, and solve system and process problems at the place where the problems occur. The strategic dimension requires that leaders set a strategy to bolster performance.

Leadership

Top 10 Quotes on Leadership are as follows (From: The 100 Best Quotes on Leadership—*Forbes*[1]):

1. A leader is best when people barely know (s)he exists, when his/(her) work is done, his/(her) aim fulfilled, they will say: we did it ourselves. — Lao Tzu
2. Where there is no vision, the people perish. — Proverbs 29:18
3. I must follow the people. Am I not their leader? — Benjamin Disraeli
4. You manage things; you lead people. — Rear Admiral Grace Murray Hopper
5. The first responsibility of a leader is to define reality. The last is to say thank you. In between, the leader is a servant. — Max DePree
6. Leadership is the capacity to translate vision into reality. — Warren Bennis
7. Lead me, follow me, or get out of my way. — General George Patton
8. Before you are a leader, success is all about growing yourself. When you become a leader, success is all about growing others. — Jack Welch
9. A leader is a dealer in hope. — Napoleon Bonaparte
10. You don't need a title to be a leader. — Multiple Attributions

Healthcare Affordability leaders often come from outside the organization, originating from places that operate with either lean, six sigma, or lean six sigma practices. Or, organizations will bring in expertise from the outside to the inside of the organization for training and development. I often refer to this practice as "Outside—Inside, Inside—Outside, Inside—Inside." That is, the desired practices and behaviors are sourced from outside the organization with the intent to implement them inside the organization. Those individuals leading the implementation of those lean and/or six sigma praxes will expose the "insiders" to methods, procedures, and customs that exist outside the organization. Some organizations choose to "go to the gembe," or the place where it's really happening to see and experience how things are done differently. Ultimately, the outside—inside and inside—outside exposure enables those on the inside to transfer the knowledge and practices to the others on the inside; hence, "inside—inside" takes place.

This is not the age old, "out of the box thinking." It's more of an approach for bringing new thinking into the box, or creating a new box and a fresh paradigm of new thinking and novel performing. For this to happen, leadership must recognize that what got them where they are, is no longer the path and approach needed for growth and advancement. Leaders must be constantly vigilant as to what is best to move the organization forward and transform the establishment in a competitive and industry relevant manner. This is the area where many leaders fail. It happens by combining, lack of recognition and realization of emerging needs and requirements, with use of the same thinking that got them to the place they are today. Prudent Affordability behaviors such as gathering customer requirements, delivering value, monitoring resources expenses, providing the right resources for delivering value, pricing products and services at a competitive and palatable level, and balancing revenue and cost for profitability, all have to be sustained.

It's leadership's role to effectively change and transform the organization as time passes. This requires adjustments and modification of strategy, structure, and systems. It also necessitates development and improvement of the people, the processes and the performance of the organization. Often, senior management and middle management may resist such change, after all their role is to manage the operation and maintain status quo in order to provide the appropriate resources to deliver value and continue to generate revenue. This natural contention between leadership and management must be addressed by those leaders at the top. Management must be developed to expect change over time, and participate in the deployment and institution of new and better methods and approaches for the business.

The primary attributes of this type of leadership can be summarized in both actions and behaviors (for additional information on Affordability Leadership, see Chapter 6 on Leadership in *Affordability: Integrating Value, Customer, and Cost for Continuous Improvement*). The five actions of leadership are; (1) Set Direction, (2) Align Resources, (3) Motivate People, (4) Communicate the Message, (5) Execute the Plan. The five behaviors of leadership are; (1) Model the Way, (2) Inspire and Shared Vision, (3) Challenge the Process, (4) Enable Other to Act, and (5) Encourage the Heart (from "Leadership Challenge," Kouzes and Posner). Combining both the five actions and five behaviors, leaders can assess themselves, as well as guide their actions and behaviors in the pursuit of Affordability.

The 10 Best Healthcare Affordability Practices at the Leadership Level:

1. Leadership is heavily involved early on…and that direct involvement early diminishes over time.
2. Methods exist for the results of activities and projects to be presented, reviewed, and readout with top leadership and key management.
3. Leadership serves to support and assist in the improvement, not dictate and demand.
4. The Direction is clear, Resource Alignment is well-defined.
5. The Strategy is known and deployed.
6. The Design and Plan for accomplishing the transformation are well communicated.
7. The Program Plan is available for all to see, understand, and participate.
8. Resource alignment fits the effort and the initiative.
9. Leadership serves to motivate the people.
10. A rhythm of performance review by leadership is established, that provides a "pulse check" for organization performance, and the resulting action by the people is a part of the culture (This operates in PDCA fashion as CAPD—Check, Act, Plan, Do).

Strategy

The strategy of an organization serves as the design and blueprint containing purpose, vision, and mission with goals and objectives for setting direction, aligning the resources, motivating the people, communicating the message and executing the plan. The strategic plan provides the details of how to accomplish the strategy. Within Healthcare Affordability, the strategy and strategic plan should address all levels of the organization (i.e., strategic, operational, tactical), and articulate how people, process, and performance play a role in accomplishing the purpose, vision, mission, goals, and objectives (For more information on constructing a strategy and strategic plan, see Chapter 6 of *Affordability: Integrating Value, Customer, and Cost for Continuous Improvement*).

All too often, a strategic plan is disjoint and disconnected from the people and the value stream. The goals and objectives, and the

Figure 4.1 Healthcare Affordability organization hierarchy.

performance metrics and measures, should deliver the current state status of execution for comparison against the goals and objectives, as well as provide data feedback on operational and tactical execution outcomes. Status and feedback provide the fundamental facts for making decisions and taking action. At the strategic level, decisions can be made to improve customer targeting, business fundamentals, process and system improvement, and the alignment, development, and motivation of the people. Most organizations are arranged in a pyramidal construct (see Figure 4.1) with three stacked general levels. The top leaders and senior managers are at the strategic and systems level. Managers are positioned at the operational and process owner level. While the people comprise the tactical, procedural, value added support, and value added delivery level.

For strategy development and deployment, input is passed down throughout the organization, and feedback is gathered passed back up through the organization. After several rounds, the strategy becomes refined, while being deployed throughout. This method is often referred to as "catch ball." Using this methodology, every individual is linked to direction and alignment, with clarity of the roles and responsibilities of everyone.

The strategic level directs and drives the operational level through strategy deployment, which in turn, guides tactical level by way of the linked goals and objectives. From the 5P Cyclical function, we know that people drive the processes, and the outcome of the processes provides performance feedback and information for the strategic level to make decisions, enact adjustments, and formulate effective course corrections.

When creating a strategy, its key critical to assess the situation and completely understand the current state of the organization and environment, before rendering a future state and designing the strategy and strategic plan. For additional detail and information, please see Chapter 6 of the foundational book *Affordability: Integrating Value, Customer, and Cost for Continuous Improvement.* The strategy design should set direction and articulate resource alignment, just as much as, provide a plan and detail for implementation and deployment.

Systems

Of course, the deployment of a strategy, and the execution of a strategic plan, requires that an organization must change and transform according to the current and emerging conditions, the course set, the direction heading, and the adjustments required to be made over time. Today, continuous change is necessary to stay competitive and continue to deliver the best value for the customer. There are five primary requirements for system to change and transform (see Figure 4.2). Vision, mission, and purpose, as well as goals and objectives, provide information for the people to understand direction and alignment as to the; where, when, why, what, how, and even who. Leadership provides; direction, alignment, motivation, communication, and execution. People need to understand the "W.I.I.F.M." (What's In It For Me), and have the skills, knowledge, capability, and competency to accomplish the plan. Processes and resources must be available to carry out the actions, activities, and work essential to attain the goals and accomplish the objectives. Finally, a plan and program must be in place to provide a detailed framework and blueprint that the organization can follow to achieve success.

When the requirements are combined, performance change can be realized. However, if one factor is missing, chaos and confusion, or

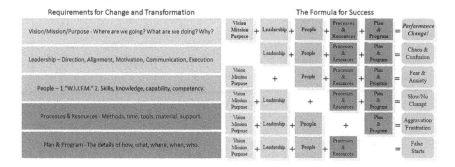

Figure 4.2 Requirements for change and transformation and the formula for success.

fear and anxiety, or slow/no change, or aggravation and frustration, or numerous false starts could be the result. Leadership must ensure that all requirements are in place and continuously monitor that all requirements are being addressed and met. Far too often a program is started with verve and vigor, and it ultimately fails because attention is not paid to maintaining emphasis on the requirements. These requirements are mandatory for a system to move forward and sustain its momentum.

Structure

In addition to strategy and system development, structural development is compulsory. A static organization structure that has been successful is not apt to change. Organizations that naturally change are dynamic and flexible. Often, such organizations are not large and strategy, purpose, vision, mission, leadership, people, resources, processes, plans, and programs are easily understood and adaptable. In larger successful organizations (those of 200 or more), change and transformation requires a structure that is beyond the static and status quo conditions that exist. Below is an illustration depicting a model using a dynamic dimension that affords the flexibility and creativity to create change and integrate the changes into the organization. It was derived from the "dual system strategy" recommended by Dr. John Kotter in the book *XLR8*[2] (Figure 4.3).

It embraces a dynamic design feature that permits change to occur independently from the successful static organization, and transformation to transpire by folding the changes back into the organization. This scheme employs a guidance team, comprised of leadership for directing the effort, a design team for designing solutions and improvements, and

A Framework for Change and Transformation

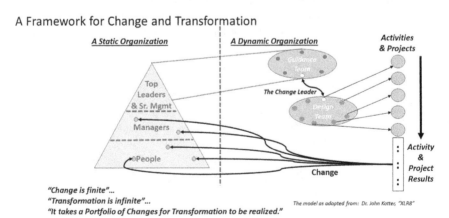

Figure 4.3 A proven framework for change and transformation. (Adopted from Dr. John Kotter, "*XLR8*", 2014, Harvard Business Review Press.)

project teams for achieving and accomplishing change. The results of the projects are folded back into the organization and change happens. Over time, this approach accumulates a portfolio of changes, and as the organization experiences numerous changes, transformation is realized.

Strategic purpose

The purpose of Healthcare Affordability is to deliver value to the patient or customer according to the requirements, needs, wants, and wishes, with the balance of cost (i.e., "balance of cost" = the balance of the expense of the resources and the revenue remuneration through competitive pricing). For care providers, it is the delivery of patient care. For the other providers in the Healthcare Enterprise, it is either the delivery of machines and devices, or services and supplies, or medication and pharmaceuticals, or insurance and payment, to care providers for the delivery of patient care. Every method of Value Delivery should be Value Stream Mapped that creates a strategic illustration of how value for the customer is provided. This strategic view enables organizations to align their resources with value added and value added support functions.

Ultimately, all value streams converge and integrate at the point of patient care and patient care delivery. The first, foremost, and primary need for Healthcare Affordability is the detailed understanding of customer requirements. With the wide variety of value streams, the amount of focus on Healthcare products and services, the scope of this endeavor is "ginormous." For every malady, sickness, condition, and disease to be detailed and described in a standard way, there should be a standard method and common tool for understanding requirements for patient care, care services and care products. The tool and method I recommend is the House of Quality (HOQ) and Quality Function Deployment (QFD) using Voice Of the Customer (V.O.C.) for requirements understanding and documentation. In its traditional usage, the relationship between patient requirements and the approaches used to meet those requirements are addressed. By studying how well patient requirements are met, how well those methods will be implemented and what processes and controls are necessary, an organization can identify and deploy improvements for care. Taking it to the next level, specific types of patient conditions can be analyzed using the HOQ tool for improvement of procedures and practices for care.

With the permission of Columbus Regional Health, in Columbus, Georgia, I've included just one example of the use of HOQ and QFD for using requirements for a particular type of patient condition (please refer to Appendix C for this section). The preparation of the model was done by Freya Gilbert, RN and Direction of Quality Management, as part of her Lean Six Sigma Black Belt IISE Certification project. The activity involved

Freya and a team of subject matter experts she engaged. The patient condition and value delivery is focused on THA (Total Hip Arthroplasty) and TKA (Total Knee Arthroplasty) procedures, commonly known as hip replacement and knee replacement. It begins with the use of THA/TKA "Requirements" or the "What's," and maps them against the "Quality Characteristics," also known as the "How's" (i.e., the How's that address the requirements). It initially provides a complete and comprehensive understanding of how both variables (i.e., Care Requirements and Quality Characteristics) are related. And, throughout the five levels, it provides a comprehensive understanding of "What" is required, and "How" the requirements will be met. Below are the details of each of the five levels of the THA/TKA HOQ:

- HOQ 1: When a hip or knee is replaced, there are seven requirements: procedure scheduled, preadmission testing, preop/holding, intraoperative period, postanesthesia care, inpatient, and discharge. How to meet those requirements involves three primary areas of treatment: clinical, equipment, and education. Evaluating the level of relationship between the requirements and the quality characteristics,

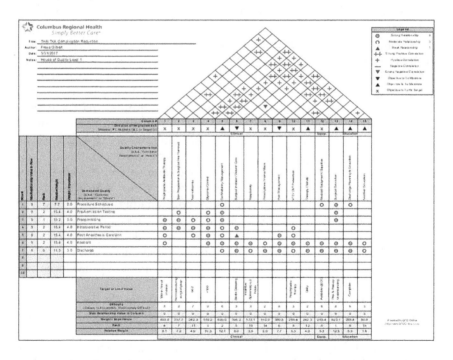

HOQ 1 The relationship between patient demanded quality and quality characteristics. (Data from http://www.qfdonline.com.)

the target amount applied for each quality characteristic, and the correlation of the quality characteristics, a comprehensive illustration of what it takes to meet the requirements is illuminated. All of the factors, weight, importance, relationships, and correlations were determined by the subject matter team by leveraging their expertise through dialogue and discussion. The outcome was specific a rank for each requirement and quality characteristics. At this level, you can see that preop/holding ranked highest, followed by preadmission testing, intraoperative, postanesthesia, and inpatient for requirements. The top three quality characteristics, or functional requirements, came out to be (1) Surgical Education, (2) Co-Morbidity Management, and (3) Glycemic Control. This information in itself provides a "one-page perspective" of the relationship of requirements with quality characteristics (aka functional requirements) for this condition, as well as the amounts and limits, and the correlations of those quality characteristics or How To's applied.

- HOQ 2: At HOQ level 2, the quality characteristics from HOQ 1 become requirements and needs to be analyzed with their functional requirements or their "How To's" applied to address those

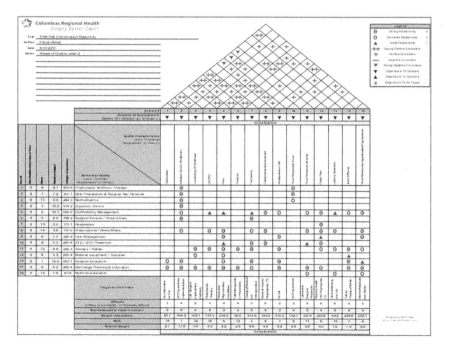

HOQ 2 The relationship between the original quality characteristics and required functions. (Data from http://www.qfdonline.com.)

requirements. As in HOQ 1, the relationships are determined, the targets and limits are declared, and how each of the quality characteristics correlate are decided. It becomes apparent at this level that the top three complication priorities are as follows: (1) Infection due to prosthesis, (2) Hematoma, and (3) Postoperative fall. But this is not the end. These How To's, Quality Characteristics or Functional Requirements must be evaluated as requirements and needs at level 3.

- HOQ 3: The complication of HOQ 2 become the requirements, to be analyzed with the How To's or functional requirements of comorbidity and financials. Once again, as in HOQ 1 and HOQ 2, the relationships, how much and correlations are determined by the evaluating team. The results are scored by the inputs and priorities appear.

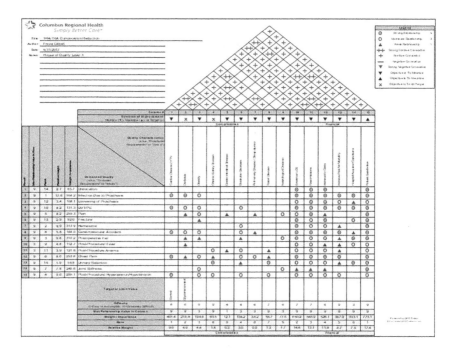

HOQ 3 The relationship between the original required functions and their how To's. (Data from http://www.qfdonline.com.)

- HOQ 4: At HOQ 4, the same procedure and activities are repeated as before yielding priorities for; Hypertension/Hypotension, Heart Disease, and Implemented Devices Causing Complications.
- HOQ 5: Finally, at level 5, the HOQ is completed and the priorities of Medication Management, Resources Patient and Family, and

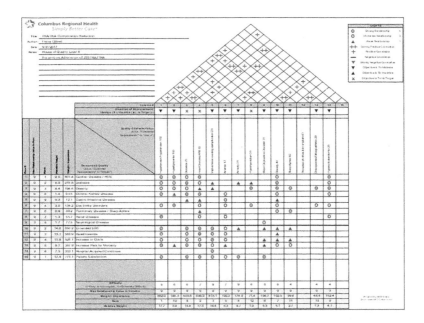

HOQ 4 The relationship between the how To's of HOQ 3 and associated characteristics. (Data from http://www.qfdonline.com.)

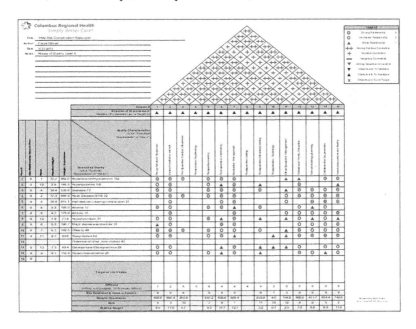

HOQ 5 The relationship between the associated characteristics and associated functions. (Data from http://www.qfdonline.com.)

Internal Medical Consultation are realized as the most critical areas, while the rest remain important to care. At each stage, the team discovered "Ah Ha's" for treatment, and a comprehensive understanding of the conditions and complication avoidance was determined.

Summary: Once completed, the team grouped and summarized each level and used the information when talking to physicians and senior leaders. This information served as a comprehensive information model to articulate the complete view of such procedures, and also as the basis for continued management, improvement and sustainment of the processes, procedures and treatments of THA/TKA patients.

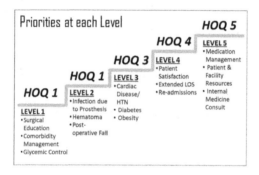

HOQ 5 The resulting priorities at each level. (Data from http://www.qfdonline.com.)

Next Steps: The data and information were then taken by the team and used for the complete evaluation of the performance of the system. This included the use of Failure Modes Effects Analysis (FMEA) for determining the best practices used, and the procedures targeted for improvement.

It must be noted this is not the traditional use and usage of the VOC, HOQ, and QFD; it does demonstrate the advanced potential for characterizing and clarifying a detailed approach to achieving a comprehensive understanding of this particular patient condition, and the specific application of care for those patients' of THA/TKA conditions and situations. I must also divulge that this methodology is quite difficult and complicated for many organizations to pursue. It takes time and effort, but the results provide a detailed, in-depth view of what it takes to deploy a standard of quality functions and standard practices throughout the organization for any value stream delivery of patient care.

Strategic value delivery

For understanding and defining strategic value delivery, a Value Stream Map (VSM) must be created. Many care providers specialize in a variety of value streams such as cardiology, oncology, pediatrics, neurology, trauma, stroke, etc. Each area of focus and specialization should be value stream mapped for the purpose of creating a unified view of each primary value stream of care flow and delivery for the institution. There has been a great deal of literature published on value stream mapping, so it would be a waste of time and script to lay out the steps and stages here. Basically, the map should be created from the patient centered perspective, mapping the steps and sequence "upstream" to the origins of the supply chain. The information, communication, work, time, and details should be included (all on one page) to provide a complete illustration and picture of how the value is delivered. The VSM serves as a "30,000 foot strategic view" of how the organization delivers and flows care. Many VSMs are accompanied by operational flowcharts and tactical procedural diagrams that explain the work required to deliver care.

The understanding purpose, patients/customers, value, and value streams are cornerstones for strategy. The pursuit of Healthcare Affordability starts with how value is delivered through compliance with patient and customer requirements. This is the proactive perspective of Affordability. The cost dimensions of expense, resources, price, and revenue is formulated from how much is required to deliver value knowing customer requirements. Strategy, strategic focus, and the strategic plan can be created from the knowledge and acknowledgment of these cornerstones.

Healthcare provider strategic focus and strategic plan

After the delivery of value is defined and mapped for deployment, it is critical for leadership to determine and address how the organization will incorporate the new strategy, system, and structure requisites. As part of every strategic plan and deployment, it is a best practice for the leadership team to create a collaborative vision, mission, direction, goals, and objectives for alignment, motivation, communication, and execution.

One outstanding example of this occurred several years ago as a part of a Lean Black Belt Certification project for June Estock, RN, Christianna Care Health System. It focused primarily on Internal Medicine and the Adult Medicine Office (AMO). It involved four executive sessions, supported by a team of facilitators and subject matter experts. The support team was involved in facilitating, researching, developing, and formulating the strategy and strategy artifacts. The executive sessions for discovery and strategy development were scheduled for two half days in September, followed by two half days in November. The entire timeframe for the activity was 4 months (September through December).

Before the first session was held, a great deal of preparation was conducted by June and the team. It involved the development of training, scheduling of the executives and other resources, and the formulation of the tools, toolsets, and techniques to be used. Although the executives were involved for only four half days, the support team was engaged throughout all 4 months. All artifacts came from executive engagement and support team development and production

The executive sessions were orchestrated in a manner to maximize the time spent by the executives and gather pertinent data and information for creating the strategy, strategic plan, and the actionable endeavors to change and advance the organization. The topics and flow were as follows:

- AMO Efficiency Initiative
- Project Focus
- Introduction to Lean, Process Improvement, PCDA
- Organization Vision
- AMO Efficiency Project Opportunity
- Value Stream Mapping
- Current State
- Value Add, Non-Value Add, Waste
- A3 Thinking
- Problem Definition
- Problem Analysis
- Identification of Metrics
- Data Collection and Analysis
- Action Plans
- Prioritization Matrix
- Sustainment
- Team Dynamics
- Change Leadership
- Strategy Elements
- Path Forward

At the culmination of the event, the course was set, the priorities were set, and the improvement activities were identified (see Figure 4.4). Its worthy

Healthcare Affordability: Focus and Plan

Figure 4.4 Elements of a Healthcare Affordability strategy and plan. (Courtesy of Christiana Care Health System, Wilmington, Delaware.)

to note at this point, Christianna Care Heath System is one of the highest ranked U.S. hospital systems in terms of quality care.

Healthcare service provider affordability strategy

Since Healthcare Affordability doesn't center on care providers alone, and to accomplish the overarching Affordability initiative, the other elements of the Healthcare Enterprise must participate in strategic efforts as well. Any organization that provides a Healthcare product or service is a candidate for Affordability. Over the past several years, I have been involved with architects and construction companies specializing in Healthcare. These businesses have a unique role to play in this enterprise. Their designs and products have a profound effect on the Healthcare providers they serve. They play a key part in providing structures and systems that support Affordability.

With one architectural firm, I've traveled the country to their various offices training them on Lean and certifying their Lean Black Belt candidates as a part of their strategic initiative to implement Lean. Each candidate has completed a project with a client and for the company, eliminating waste, improving processes, and increasing profitability. The net result has decreased costs and improved process, while accelerating momentum and increasing quality. Often, the solutions involved the organization partnered with the client to create a collaborative solution with an enhanced level of service care.

With another firm, I spent several weeks delivering a Lean Green Belt program for all personnel in an effort to support their Lean strategic initiative. This included personnel from all disciplines across all of their core offices. Although initially, they wanted the knowledge and capability to serve the customer, their enlightened discovery appeared when they realized, Lean practices could serve them internally, just as much as it could be used to serve the client. Their efforts from that point forward had a dual focus of company and client.

My final example has occurred more recently. After delivering a concentrated, comprehensive program for the entire establishment, I partnered and participated with them on a project involving one of their key customers. After this firm experienced years of success with this customer, the customer, under new management, chose to contract a Lean consulting firm for interior re-design of a few of their critical care areas, with dismal results. The hospital decided to reconvene the relationship with my client and pursue a new Lean design for their food and nutrition service. After good initial results, the project is approaching design finalization and implementation.

Although these three examples are of one small segment of the service and suppliers Healthcare Enterprise element, they demonstrate how Affordability can be implemented in any portion of the provider sector. And beyond just services and supply, the possibilities transgress into the machine and device provider sector, the medicine and pharmaceutical provider sector, and the insurance and payment providers. Any business that touches Healthcare can benefit from adopting Affordability. Strategically, all of these efforts are focused on increasing performance. Increasing performance internally and externally with customers and suppliers. Increasing performance in the areas of process speed, product and service quality, and cost reduction with increased profitability and customer demand.

Performance

In order to realize strategic success, the execution outcomes of the strategy and strategic plan must be measured. To use a common quote, "How do you know if you're winning or losing if you don't keep score?" And, to quote several business leader gurus, "What gets measured gets done." Every successful strategy initiative includes a scorecard used to communicate accomplishments and identify opportunities for improvement, as well as identify initiatives, goals, and objectives. Kaplan and Norton's Balanced Scorecard (the book: *The Balanced Scorecard*, 1996[3]) provides a good framework for the creation, development, and deployment of strategic measures that set direction and provide alignment. The example below is a rendering of a balanced scorecard for some regional medical center.

YRMC ("Your Regional Medical Center") balanced scorecard

Vision	Quality of life and health for everyone in our community			
Mission	To be the primary care provider in our region			
Strategic priority	Patient outcomes	Quality and safety of care		Healthy community
Strategic results	One stop care for all patients in the region.	The state benchmark for quality and safety.		Healthcare partnership with the community, meeting all requirements.
Category	Strategy	Measures	Targets	Initiatives
Customer	• Patient satisfaction	• Patient survey	• 92% overall	• Patient satisfaction survey program
	• Accessibility to care	• Ontime and available healthcare for patients	• 10% increase	• Performance improvement program
	• Quality and safety of care	• CMS, joint commission, NPSG, IOM, AHPQ	• Per set goal	• Quality and safety improvement program
Business	• Increase revenue	• Revenue per target value stream	• 12% per year	• Increase demand through affordability
	• Decrease cost	• Operating cost	• 8% per year	• Waste elimination initiative
	• Increase profit	• Net profit	• 5% per year	• Improve revenue cycle process
Process	• Value added processes	• Per process: time, quality, cost	• Per process goal	• Affordability
	• Value added support processes	• Per process: time, quality, cost	• Per process goal	• Affordability
	• Supplier processes	• Per process: time, quality, cost	• 100% involved	• Supplier affordability training and development
People	• Employee satisfaction	• Employee survey	• 87% overall	• Quarterly employee satisfaction survey
	• Knowledge and skills	• Individual employee development plans	• 92% of goal	• Competency and capability initiative
	• Technology	• Individual technology and tool competency	• Per set goal	• Technology proficiency program

This scorecard provides a one-page diagram of the vision, mission, strategic priorities, strategic results focus, and the metrics and measures for the four key strategic areas: Customer, Business, Process, and People. It is a template to use for illustrating the strategy areas, the measures of success, the targets or goals being pursued, and the initiatives being utilized. Of course, behind the scorecard is the strategy, strategic plan, and the artifacts that explain the required actions and activities required.

The strategic measures should be linked to the operational and tactical measures of People → Process → Performance. Over the past 30 years, I've researched and identified the areas of measures that best articulate success and failure. Below are the basic foundational measures for Affordability at the operational and tactical levels:

Qualitative (The People Dimension)

- Customers
- People
- Suppliers

Quantitative (The Process Dimension)

- Time
- Quality
- Cost

With the complete understanding of the integration of value, customer and cost, in concert with the strategic, operational and tactical areas, combined with people, process and performance, a comprehensive scorecard of success can be implemented in any Healthcare Enterprise participant.

The strategic dimension begins that cascade down through the organization from leadership to the value stream delivery function. As in the "catch ball" process, from leadership, to senior management, to operational management, to the tactics and the people delivering value to the customer, each level and linkage is critical. And also as critical is the return pathway of performance and communication, from the people, to the management, to the senior management, to the top leadership. It's not a "tops down" approach, nor a "bottoms up" approach, but in context, a dual directional, comprehensive approach.

References

1. *Forbes*, The 100 Best Quotes on Leadership, 2012, https://www.forbes.com/sites/kevinkruse/2012/10/16/quotes-on-leadership/#780afbad2feb.
2. John Kotter, *XLR8*, 2014, Harvard Business Review Press.
3. Robert S. Kaplan and David P. Norton. *The Balanced Scorecard: Translating Strategy into Action*, 1996, Harvard Business School Press.

chapter five

The operational perspective

At the operational level, process is the priority. The systems of the organization are comprised of linked and integrated processes that contain procedures and methodologies that utilize the resources of the establishment. For any company or corporation to operate effectively and efficiently, the strategy, structure, and systems must function by way of robust, durable, and proficient processes. Dr. W. Edwards Deming on process:

- If you can't describe what you are doing as a process, you don't know what you're doing.
- Eighty-five percent of the reasons for failure are deficiencies in the systems and process rather than the employee. The role of management is to change the process rather than badgering individuals to do better.
- We should work on process, not the outcome of our processes.
- The process will produce what the process is designed to produce.

Although a great amount of functional activities are at the operational level in every organization, the primary ingredient for Healthcare Affordability success is process. Every division, department, or branch of the establishment functions by way of a collection of work processes. When an enterprise is successful, the business and value systems and their processes deliver value to the customer in an efficient, effective, more affordable manner. The way to improvement, and the best method to sustain competitiveness, is to continually improve processes. This impacts the speed, quality, and Affordability, thereby guaranteeing customer satisfaction and competitiveness.

Process

A process is a group of related activities and actions the transform one or more kinds of supplied inputs into output that is of value to the customer. Process can occur sequentially, or they can operate in parallel within a system. A system is comprised of coordinated and integrated processes. Business Systems (e.g., Sales and Marketing), Value Systems (e.g., Customer Product and Service Production, and Delivery), and Value-Added Support Systems (e.g., Supply Chain and Maintenance) contain processes designed to accomplish the functions and tasks of each area of the organization. In Healthcare Affordability, in a hospital, the processes of the Business Systems (e.g., Accounting, Purchasing), the processes of the Value Systems (e.g., Emergency, Cardiology), and the processes of the Value-Added Support Systems (e.g., Housekeeping, Registration) are choreographed to deliver care.

In its basic form, a process is comprised of five basic elements. The sequence of elements occurs when a supplier, or suppliers, provides input, or inputs, to the process that delivers output to the target customer or customers. Surgery, for example, requires tools, materials, supplies, and resources, provided by suppliers, to perform the procedures necessary to care for, and heal, the patient. The output of the surgery meets the requirements of the patient needs.

The detailed anatomy of a process (see Figure 5.1) contains the basic process components (i.e., Supplier, Input, Process, Output, Customer), as well as the factors that provide information for improvement and intelligence for management. Three keys in the management of an outstanding process are requirements, process measurement, and communication. The process itself should be provided with all needed resources to deliver value to the customer per the customer requirements, which includes quality input from suppliers per the process' requirements. The internal requirements for process operation should be documented, diagrammed, and illustrated in a standard manner, along with all details of the supplier, input, output, and customer components.

Operations management and process owners should review the status of operation of the process with a regular rhythm (e.g., weekly). The people involved in executing the process should review process performance daily and make improvement efforts an ongoing discipline. Process

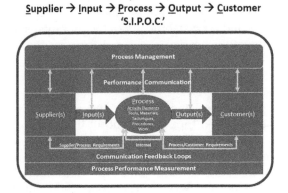

Figure 5.1 SIPOC.

performance review should include process status, current performance compared to goal or objective, opportunities for improvement, status of improvement efforts, resource needs for improvement, recognition, and celebration of improvement success (Note: Process Performance Review should not take more than 15 minutes.).

In an Affordability world, top leadership assures alignment throughout the organization for all system and functions, while management is responsible for owning and managing processes and operations. The people are positioned to perform the work (business, value added, and value-added support) and improve the processes in which they are involved, as well as assist in process improvement for processes where they have functionality competence or knowledge.

Systems and processes

The systems and processes of an organization is how work gets done and value is delivered to the customer. Whether the systems and processes are unstable, undocumented and informal, or quite mature, standardized, and well managed, they exist in some form or other. It is through these systems and processes that the value stream is able to flow products and services from supplier to the customer. Healthcare Affordability is attained through the practice continual improvement. Whether it be care providers or their providers, the pursuit of Healthcare Affordability requires continuous improvement of systems and processes.

Outpatient surgery is an example of a commonplace Value Stream System employed in Healthcare (common outpatient surgeries include cataract treatment, endoscopy, colonoscopy, injections, biopsies, labs, excisions/removals, physical repairs). Typical steps, stages, and processes for this system involve registration, preparation, procedure, recovery, and discharge. Although each process is an entity unto itself, when sequenced and integrated the system qualifies as a Value Stream. By improving the system, the value stream of delivery is improved.

Revenue cycle is the customary Business System in Healthcare institutions that collects the revenue and functions regularly manage profitability. The usual steps, stages, and processes for this system involve:

- A Patient requires a healthcare service, and it is initiated by a referral from a physician.
- The Patient's appointment for the service is scheduled.
- The service is rendered by the organization and the patient is released or discharged.
- The insurance and payment provider is billed for the service.
- The service is paid for, and the revenue cycle is complete.

This internal business system is comprised of processes that support the business. The value stream of delivery is included, the revenue is collected. This system serves the business of the organization.

By focusing on Value Streams (including systems that support those Value Streams) and Business Systems, the operational portion of the organization can maintain, sustain, and improve performance. It is not as important to focus on results and outcomes, as it is to focus on improvements and enhancements to the systems and processes. The system and process advancements will yield better results and outcomes. Just like Dr. Deming said, "Management by results is like driving a car by looking in the rear view mirror."[1] To keep the car on the road and headed toward the goal, attention must be kept on adjusting and maintaining the car's heading toward the proper direction by vigilance and course correction. The results are ultimately realized when the goal is reached, and the car arrives at its destination. In other words, operate the process, review the performance, and ever improve the process, to reach the goal.

People and performance

People drive the process. Even when the process is automated, the performance must be reviewed with regular rhythm and rigor, and automation adjustments or improvements must be made when the process is not operating within design, requirement, tolerance or standards in terms of speed, quality and cost, and of course, customer satisfaction. If improvements are needed, the problems of the process need to be fixed by the people. The people who own, know, and work with the process regularly. If subject matter expertise is required, the organization should provide that expertise for process problem-solving.

Today, across the Healthcare Enterprise, involved organizations go about improving processes and performance in a variety of ways. Some have a process improvement ("PI" department), some hire consultants, some use their engineering resources, others make managers find their own methods, and others use a hybrid of approaches (not all methods are covered in this list). Those using the "people solving their own process problems" approach are experiencing great gains in performance and profitability. In addition, this particular approach stimulates motivation and encourages teamwork.

To improve customer satisfaction, improve the processes and operations that deliver value. To improve the business, improve processes and operations that involve people, the flow of money, necessary materials and services, information and technology, sales and marketing, risk and legal, conformance and compliance, and above all, strategy. To improve the workforce, improve processes and operations in addition to establishing and fostering meaning (purpose), mastery (competency), and membership (teamwork) for every employee or associate. To grow, stay competitive and stay in business, improve processes and operations. The heart of Healthcare Affordability is about improving the processes and operation through; motivating people who improve the processes and the result is increased performance.

Process and operational improvements

Process and operational improvement are key to operational success. Dr. Deming, renowned for his work in process improvement and process management, is the best person to provide principles for improvement. Five of his principles that have a big impact in this area are as follows:

1. Quality Improvement is the science of Process Management.
2. For Quality Control in Healthcare, if you cannot measure it—you cannot improve it.
3. Managed care means managing the processes of care, not managing physicians and nurses.
4. The right data in the right format, at the right time, in the right hands are most important.
5. Engage the clinicians (the "smart cogs") of Healthcare. The frontline people who know the process best. Understand what is necessary to improve, and the ones who actually own care.

Dr. John Haughom[2], Senior Advisor, Health Catalyst says, "Applying these key Deming principles to healthcare process improvement can help every healthcare organization show the workforce why change is necessary, what they need to understand to participate in meaningful change, and what success will ultimately look like." In other words, institute process management ... measure the process to gain a sense of the actual performance status ... manage the process, not the people ... utilize the right data, at the right time, in the hands of the people who can improve the process ... leverage the clinicians, who are highly committed, intelligent and capable.

It is management's responsibility to own and improve the process. The people are improved by education, training, and development. Management must develop a systematic approach to problem-solving that includes and involves the people who know the process best, and the talent required to develop a solution to the problem in the process. Management must create an environment that finds problems, solves problems, and recognizes the victories and successes.

Management must review process performance status on a regular basis, using a standard rhythm, using a review assessment of: process performance status, status as compared to goal, improvement opportunities to pursue, improvement projects and status, needs and requirements for improvement, and lastly, but perhaps most important, celebration of successes and recognition of those who accomplished improvement.

The pursuit of process maturity

The view of an excellent system and a best class approach was explained in the last section. I have found that it is most often the case that process maturity for many organizations is underdeveloped. Quite often I've discovered, operational processes are undocumented or poorly documented, and often, when they are present, they are disobeyed and not a standard. Even when the processes have some level of standardization, they are commonly fraught with waste. I have developed a "rule of thumb" for process maturity at a high level. Maturity can be observed as follows;

- Unstable: Processes and process thinking is absent in the organization. Standard for work are "tribal knowledge." The goal is to Stabilize—Document all Processes (Use Documents, Charts, Maps, and Illustrations)
- Nonstandard: Although there is process documentation present for some or all processes, work is being accomplished by way of personal knowledge or individual experience. Measurement may be in place but not utilized to manage the processes. The goal is to Standardize—Use Lean (Create Flow, Eliminate Waste)
- Unmaintained: The processes and standards are established, but there's not an effort to sustain the effort. Processes are being measured, but the measures are not being used to identify opportunities and improve performance. The goal is to Sustain—Use Six Sigma (Increase Performance, Reduce Variation)
- Sustained and Maintained: Process management is in place. Processes are being improved by those working in the process, with the assistance of subject matter experts when required. Continual Process and Operational Improvement

At each level, a goal and initiative can be applied to increase speed, improve quality and lower cost. Applying quality practices, Lean methods, and Six Sigma methodology would increase overall performance and achieve customer satisfaction.

About 30 years ago I had my first in depth exposure to process management, quality initiatives, and performance improvement while working for NCR and AT&T's Bell Labs. I was fortunate enough to be provided with opportunities with various organizations applying Lean and Six Sigma. I became a "Deming Disciple" and realized, if you use the people to improve the process, performance naturally increases. In fact, the people involved in the process discover better approaches to improvement than other professionals assigned from the Process Improvement Department. With this knowledge, experience, and exposure to a variety of companies, from a multitude of industries, the pursuit of performance improvement can be accomplished by engaging the people in process improvement and process management and involving management in process performance on a regular basis. Specific to the Healthcare Enterprise, I have several examples below that demonstrate the possibilities and illustrate the results of instituting Healthcare Affordability.

Pediatric surgery

While I was coaching and mentoring Isaac Mitchell, a Lean Black Belt Candidate working at East Tennessee Children's Hospital, he engaged teams of his organization with the improvement of several processes. The theme of his effort was focused on Pediatric Surgery, and by including other portions of the organization, they realized substantial improvement.

By mapping Pediatric Surgery, he was able to establish a baseline of Value Add = 42%, Surgery Lead Time (Start to Finish Average) = 102 minutes, Patient Lead Time (Time Separated from Parents to Time Reconnected with Parents) = 247 minutes. After removing a great deal of waste in the system and processes, profound results were realized; Value Add = 49%, Surgery Lead Time = 33 minutes, Patient Lead Time = 67 minutes. Note: Overall Improvement in Time, Quality, Cost.

In addition, he applied Lean techniques to improvement; workplace organization, material and supply flow, use and utilization of effective visuals, problem-solving using A3 thinking, and strategic deployment of Hoshin Kanri and Metric Management Boards. Over the years, his approach and methods spread throughout the organization.

Readmissions

At a Georgia regional medical center, a readmission reduction project was launched for several reasons: Hospital reimbursement was being reduced and they were being penalized for readmissions, families and patients have lower satisfaction when readmission occurs, and valuable services were being consumed when patients are readmitted, thereby decreasing accessibility to care for others.

The duration of this project spanned 33 months, beginning in June 2010 (Baseline) and completed in March 2013. There were several key initiatives employed including: use of a roles and responsibilities matrix (RASCIN) for accountability and alignment, process identification and management, heavy inclusion of nurses for problem-solving,

"Teach Back" for improving Patient-Provider Communication, Improved Discharge Folder, Value Stream Mapping and Process Mapping, Lean Waste Elimination, Revised Physician Order Set, Daily Unit Huddles, and Community Healthcare Connection Collaboration (Involving Nursing Homes and Home Health Services to assure Patient handoff quality and continuity of care).

Outcomes and results included; Readmission Rate Improvement (see chart below), Reduction of Length of Stay by 60% (see chart below), 2013 Revenue Loss Avoidance of $864,271.00, increase in Customer Satisfaction and Quality.

Pneumonia Readmission

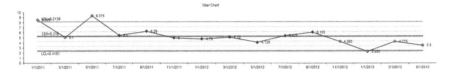

Length of Stay

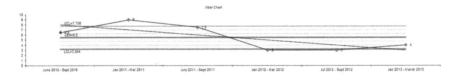

Lessons Learned

- Keep to one process at a time for rapid cycle improvement
- Need consistency of team for efficiencies and effectiveness
  - Lost Chief Nurse Officer, Care Manager, and Quality Director during various stages of the initiative
  - Nurse Manager out on Medical Leave of Absence

Discharge

The Problem Statement: the volume of patients waiting to be transferred to the Medical/Surgical Units is highest at 11 am. This is secondary to an increased number of patients in PACU awaiting surgical beds and the ED operating at a high-volume holding patients awaiting medical or surgical beds. The above often cause a bottleneck which leads to delays in patient flow. Both the PACU and ED need bed placement on the units so they can accommodate more patients.

The Goal: In fiscal year 2014, the unit discharged 7.32% patients (mean) before 11am, and years prior were all less than 8%. In fiscal year 2015

YTD (year to date, Q2) a modest improvement of 8.08% (mean) was mea-sured. The goal was to have the unit discharging more than 11.5% of their patients by 11am monthly. The use of Lean Six Sigma to Solve Problems and Increase Process Performance was prescribed as the resolve.

The Treatment: Process mapping and Process Management, Root Cause Analysis, Resolving Major Barriers (Late Discharge Orders by Physicians, RN Workload), Operational Major Improvements (Contract revisions to incentivize physicians to complete discharge orders early, Individual recognition to nurse who help get the patient out by 11 am, Feedback to unit leadership on discharge compliance of the week prior.).

The Results (Note: Improvements Implemented June 15, 2015):

FYTD	Mean (%)	Median (%)
FY11	7.58	8.99
FY12	7.64	7.70
FY13	6.85	7.26
FY 14	7.32	5.13
FY15	**10.00**	**9.32**
FY16-YTD	**16.84**	**16.64**

Hospital design and construction

The design and layout of many hospitals today contain a great deal of waste, especially in the areas of waiting, motion, transportation, and defects. This initiative includes a major architectural firm, and a lead-ing research and treatment center for cancer, diabetes, and other life-threatening diseases, for the purpose of improving the pathology laboratory service group. A Clinical Lab Improvement Team was created to streamline efficiencies, reduce redundancies, Lean out processes, and configure ideal space for lab services. Below summarizes the challenges, direction, goals, content of the approach, and results.

Challenges to Address

- Multiple lab directors with individual visions and interests
- No consensus on what should move to the DOC
- Existing Lab space in poor condition and not designed for lab functions
- DOC space is too small to account for all clinical lab services moving
- No Empty chair (space) on campus
- Long tenured staff use to existing workflows
- Duplication of processes and services

Vision and Strategy

Vision: To be an efficient, comprehensive Clinical Lab providing outstanding patient care, and conducting innovative research focused on eliminating cancer, diabetes, and other serious illnesses.

Strategy: Utilize LEAN methodologies to improve efficiencies, eliminate redundancies in workflow and space, and create the ideal flow for clinical lab specimens.

Project Goals:

- Increase Throughput 10%
- Reduce FTE 5%
- Reduce Square footage 20%
- Reduce time to build 5%
- Reduce construction cost
- Increase staff satisfaction
- Minimize disruption to Operations
- Align Vision for DOC building

Order of Lean Implementation Tools and Techniques:

- Start with Voice of the Customer
- A3 Problem-Solving
- Team Kaizen Events
- Current State → Future State
- Lean Design for Space
- Decision Matrices
- People Solutions at the Gembe
- Promote One Piece Flow
- Avoid Overburden
- Minimize eight Wastes of: Overproduction, Inventory, Defect, Motion, Over Processing, Waiting, Transportation, and Skill Abuse
- Reduce Variation
- Transition from Push to a Pull System

Goals compared to Outcomes and Results

- Increase Throughput 10%: Results: Increase throughput by 8% for CMDL as measured in time (553 to 507 minutes) and 13% for Hematopathology as measured in time (335 to 290 minutes).
- Reduce FTE 5%: Decrease by 2 FTE's = $110,000/FTE/Year × 2 = $220,000 annualized and ongoing savings.

- Reduce Square footage 20%: Decrease SF by 1,085 SF of renovation space needed.
- Reduce time to build 5%: Shorten Construction Schedule by 20 days.
- Reduce construction cost: Decrease construction Cost by $1,106,700.
- Increase staff satisfaction: Decreased Staff Walking distance by 73 minutes.
- Minimize disruption to Operations: Reduced Phasing by two phases in Northwest renovation.
- Align Vision for DOC building: Users Buy in through Process.

Speed of service

During her pursuit of Lean Black Belt Certification, Lauren Ford, MHA, Operations Manager, Patient Flow Center, Emory University Hospital, embarked on a mission through the EUH Patient Flow Center to increase MRI Access. This project was facilitated by the EUH Patient Flow Center Team and centered on MRI Exam Turnaround Time.

Using Neurosurgery, and its Current Value Stream Map, the need for access and turnaround time improvement for MRI was identified. This launched the MRI Turnaround Project. DMAIC was utilized for ordering and sequencing activities: Define (Project Charter, MRI Value Stream Map), Measure (Time Studies, Clinical Data Warehouse Extracts), Analyze (Root Cause "Fishbone" Diagram, Data Charts and Target Waste Understood), Improve (MRI Technologist Standard Work), Control (TOC Adoption, Tracking of Metric Results). In addition, 5S was used to create a Standard Workplace for all three MRI locations.

Achievements: The following accomplishments were realized:

1. Engaged a Clinical Support multidisciplinary team led by the Front Line supervisor.
2. Trained MRI team on Lean Principles and how to do 5S.
3. Improved Average MRI turnaround time from 27.3 to 14.8 hours in 2 months
4. Identified over $15K of expired supplies within MRI.
5. Centralized supply storage and reduced inventory for the department by identifying 95% of supplies as next day delivery.

Lessons Learned

1. Critical to have an engaged front line leader, cannot be pushed through by "consulting" expertise alone (PFC).
2. Understanding of what flexibility exists regarding existing parameters such as staffing levels, existing case load, or working during business hours of the unit.

3. Need to be able to clearly explain the benefit of inventory management and its effect on reduced waste to staff, both in physical inventory and time spent.

Lean Six Sigma deployment

Finally, in this group of examples, I'd like to showcase a sizable Healthcare
System located on the east coast of central Florida who has been putting in
place a corporate initiative to deploy a Lean Six Sigma Mindset and establish Lean Six Sigma Practices throughout the organization. It is focused on
training and the development of personnel, and the application of Lean
and Six Sigma within the organization. The tools and techniques included
(Root Cause Analysis by using the Ishikawa Diagram with the 5 Why
Analysis and Pareto Diagrams, 5S for Workplace organization and standardization, DMAIC (and the associated Six Sigma Tools) as the project
problem-solving methodology, PDCA, and a customized version of A3 as
the Team Problem-Solving framework) supply the toolbox for the pursuit
of continual improvement.

This strategic deployment puts a foundation in place for instituting
Healthcare Affordability. Because of the size of the footprint of this organization, the deployment is still taking place, and improvement yields
are beginning to accrue. This is a prime example of the requirement of
time and effort necessary to implement, and the need for managers to
exhibit participation, patience, and commitment. It is often the case that
a 3–5 year horizon is required for the first phase, and always the case
that operational managers need to provide ongoing support, engagement,
and involvement. With thousands of employees in the organization, the
paradigm change and culture shift that's required in this case requires a
monumental amount of energy and effort. Change the people, change the
system, and change the performance.

Common themes of process and
operational improvement

The 6 case studies provided in this chapter contain a great number of
common themes. These mutual topics, shared across these examples,
serve to frame actions and behaviors required to pursue and accomplish
Affordability. This list represents a great deal of the common areas of
focus for organizations pursuing Affordability. However, it is not the
entire list of all elements of the pursuit of Healthcare Affordability. It
does represent a good portion of the ones that are most critical. Below
are two dozen+1 topics that are commonly found in Affordability
endeavors;

1. The Effort is Customer Focused and Centered! A Universal Focal Point for Affordability Efforts!
2. Opportunity Finding: The people and teams are put to task to find opportunities to apply problem-solving efforts. The fuel for continual improvement is problem-solving, and for that, opportunities must be discovered, addressed, and solved.
3. Goals: Aims and targets are established to set direction and provide a path for resource alignment. These goals are often framed in a "SMART" (Specific, Measurable, Attainable/Achievable, Realistic/Relevant/Results Focused, Timely/Time Bound) manner. No matter how it's "spelled," the goals should be established such that the organization gets the direction, alignment, motivation, and message to pursue a successful execution of the plan. Of course, it makes sense to say, "If there's no AIM for the organization, the people's work is AIMLESS."
4. Engaging the People: The people that work in the process are the best resources to employ for "fixing" the problems in the process. People improving their own workplace has been found to be motivational. The use of the intellect and the mind of the people creates a loyalty and constancy for the workforce. People develop a meaning in work when they're included in problem-solving.
5. Training, Learning and Developing People through Improvement Implementations: Knowledge and capability as applied, proves to be one of the best methods for creating capability and retention. Sometimes called: "Experiential Learning."
6. People Solving Process Problems: Instead of reactive firefighting, root cause analysis and process problem-solving is instituted using the resources of the people working within the process. Subject matter experts are called upon when specific expertise is needed.
7. Teamwork: Cooperation, coordination, and collaborative are three key attributes of teamwork. Healthcare Affordability uses teamwork as the approach for solving most problems and resolving most challenges. When an individual can quickly resolve and issue, that's often called a "Just-Do-It", and when appropriate, it should be encouraged. However, many process problems are complicated and difficult, requiring a team of people with multiple capabilities and capacities. Often said, "Two or more heads are often better than one."
8. Management's Primary Operational Focus is on System Performance and Process Improvement: The management of the value stream and its delivery of products and services is established as the manager's chief function and concern.
9. Elimination of Chaos: Stabilization of the system by eliminating chaotic activities such as firefighting, rework and wasted motion

and actions. Processes should be mapped and documented and reviewed by both management and the workforce for effectiveness. Chaotic opportunities should be identified and removed.

10. Institution of Consistency: By standardizing processes, normalizing work procedures, and removing wasteful activity, consistency can be accomplished. This is achieved by eliminating waste and creating flow. Lean is best applied for this initiative.

11. Establishment of Predictability: This provides the platform for sustainment. With predictability, Six Sigma methods can best be utilized. Increase in process performance and reduction of variation can be achieved with stable, standardized, sustained processes.

12. Pride of Work is Fostered: The meaning of the work being done is a sense of pride in the workforce. A path for the mastery of skills and capability is available for those performing the work. Membership and inclusion in team-based problem-solving and improvement efforts, is part of the regular way of doing business.

13. Motivation, Recognition and Reward: The driver of process improvement is people. For maintaining and sustaining improvement, motivation, recognition, and reward is paramount. The best motivation and recognition comes from customers, respected leaders, and teammates.

14. Measure, Review, and Continuously Improve: Management should put systems in place that are measured and reviewed for; status, performance against goal, improvement efforts, accomplishments, and needs. A rhythm should be put in place for regular reviews. Teams can review their own work daily, but managers need to review performance at least weekly.

15. Qualitative and Quantitative Measures: Both qualitative measures (such as customers, people, suppliers) should be instituted, along with quantitative measures (such as time, quality, and cost).

16. Lean—Eliminate Waste and Improve Flow: Removing waste from a system and increasing flow is a primary focus for operational improvement. Once systems are somewhat stable, an effort can be incorporated that focuses on seeing and eliminating waste, with initiatives to increase speed and flow of the processes.

17. Six Sigma—Reduce Variation and Increase Process Performance: When the processes become predictable, six sigma can be used to reduce variation and increase performance.

18. Improve Operational Performance through Improving Process Performance: Overall operation performance in terms of speed, quality, and cost improves as a result of process performance.

19. Observation: Management must serve to observe, coach, and mentor the people driving the process. Management must also teach the people to see the waste in the process and identify the opportunities for improvement.

20. Monitor, Manage, Encourage: This is managements role and responsibility. Monitor operations and performance. Manage the processes. Encourage the people to improve the processes.
21. Review Rhythm—People, Process, Performance: Management's three areas of review are people (the status of the human resources involved in the processes), process (performance status and improvement efforts), performance of the team in terms of all established metrics and measures.
22. Standardize through documentation, maps, visuals and the implementation of standard procedures: Standard information and standard methods are critical to consistency and reliability of the process. All method that apply should be used to establish standards. Compliance with standards is one of the three key dimensions of process quality.
23. Systemwide Operational Analysis: Managers and Senior Managers must hold regular reviews of operational performance, including the analysis of systems and processes.
24. Systems approach to Process Management, Process Improvement: By eliminating "silos" and instituting cross functional, linked systems, management should create collaborative environments for flow speed, quality excellence, and cost optimization.
25. All Operational Systems are linked to Overall Organization Enterprise Goals: Top Leadership goals for customers, business, process and people should be linked by specific metrics and measures throughout the organization and down to every individual.

Summary

The Operational Perspective is the Manager's realm. Managers manage things, but they must lead people to operate in a standard manner, improve processes, and to participate in changing the system to stay competitive and deliver what the customer needs, wants, and requires.

For Affordability, at the Operational Level, Process Management is the Manager's Fundamental Function. Work and Value Delivery is accomplished by way of integrated processes. Managers should focus on, and spend ample time dedicated to process management. It's within the management of the process that customer requirements drive value delivery.

Process Thinking is the Manager's Obligation. When problems occur, it's not usually the person, but a break down in the process. Managers ask, "Where did the process break down, and who can we use to fix it?" Rather than, "Who did it, and what's the problem?"

Performance and Success is based on Process Outcomes. By managing and improving the process, accomplishments, and results are realized. Focusing on outcomes does not improve performance. Focusing on improvement with increase outcomes and success is the result. A constancy of the purpose of process management and improvement must be established and sustained.

- Key Points
  - Celebrate Successes
  - Recognize Accomplishment
  - People are the best source for Solving the Problems in the Process
  - Train and Develop People, Coach, and Mentor for Developing Knowledge and Capability
  - Performance Improvement is not a Down-Sizing Strategy
  - Redeploy Available Resources to Support System Requirements
  - Establish Meaning, Develop Mastery, Promote Membership (Teamwork)
  - Management is about Things, Leadership is about People. Lead People, Manage Processes
  - Communication, Transparency, and Execution is the Manager's Mantra
  - Respect, Integrity, and Trust are the Excellent Manager's Values
- A List of 10 Best Practices at the Operational and Management Level
  - Management is Process Focused for Performance and Improvement
  - Management has "Adopted the Affordability Philosophy"
  - Management Supports the Initiatives related to Affordability
  - Management Trains, Develops, Coaches, and Mentors the People
  - Management knows the process is the problem, and fixing the process improves performance.
  - Management knows that the people solve the problems in the process best.
  - Management knows that they are there to own the process, manage processes, and lead people.
  - Management understands that process performance is their measure of performance.
  - Management owns the process and is accountable for process performance.
  - Management knows that they are the key link between strategy and value delivery.

References

1. W. Edwards Deming, 2018, Quote by W. Edwards Deming, http://quotes. deming.org/authors/W._Edwards_Deming/quote/4978.
2. John Haughom, Five Deming Principles That Help Healthcare Process Improvement, 2018, https://www.healthcatalyst.com/insights/5-Deming-Principles-For-Healthcare-Process-Improvement.

chapter six

The tactical reality

The tactical level is where the value delivery occurs, and the people perform the value work.

- Dr. W. Edwards Deming on People[1]:
  - The People aren't the Problem.
  - The Process is broke.
  - Fix the Process.
  - Put the People to work fixing the Process.
- Margaret Wheatley's Ten Principles for Creating Healthy Communities[2]:
  - People Support what they Create.
  - People Act Responsibly when they Care.
  - Conversation Is the Way Human Beings Have Always Thought.
  - To Change the Conversation, Change Who is in It.
  - Expect Leaders to come from Anywhere.
  - We Focus On What Works and It Releases Our Creative Energy.
  - The Wisdom resides within Us.
  - Everything is a failure in the middle (When things are falling apart, what is done?).
  - Humans can handle anything, as long as we're together (teamwork).
  - Generosity, Forgiveness, Love (Values and Principles).

The people, The work, The value

The tactical reality exists at the "People Level." The people deliver the value to the customer as designed by the process per the customer requirements. The outcome of the work becomes the performance. If indeed, the process delivers what the customer requires, wants, needs, and wishes for, then satisfaction, if not delight, is the result. In addition to meeting requirements, ontime, quality, and cost are all part of the satisfaction equation. The system operates as follows, and the tactical consequence is value delivery. The Value Delivery Sequence: People → Process → Performance

The performance results include both qualitative factors (i.e., customer) and qualitative factors (i.e., requirements, time, quality, cost).

This is also known as "The Bottom Line." The interpretation of the bottom line usually equates to a financial value, but in Affordability, the bottom line is both a qualitative and a quantitative outcome. Using the Balanced Scorecard (i.e., Customer, Business, Process, People) at the corporate level, and linking the operational level and the tactical level to it throughout the organization, the point of Value Delivery is where performance is best determined. So this linkage, if viewed correctly, starts at "the bottom" and drives success to the top. The Performance Linkage Sequence: Tactical → Operational → Strategic.

Healthcare Affordability delivers value per customer requirements, while balancing expenses for the value delivery with the resources required and accumulates revenue from the competitive price charged for that value delivery. The measure of success is gained from the qualitative and quantitative results at the tactical level, which is integrated within the systems at the operational level and accumulated at the strategic level. Therefore, people are the drivers of the process and performance is realized from the results of process output. In order for strategic goals to be met, systems within operations and the processes driven by the people should be constantly measured, monitored, and improved. With the people performing the work through the processes in place, the true bottom line improves when people are motivated to improve the processes and increase the performance.

Pursue Healthcare Affordability: motivate people → improve processes → increase performance

Such a pursuit is not some fleeting initiative to be chased until a new more alluring initiative appears. It's a quest to be instituted and inculcated, in a way to become "the way business is done." Affordability is designed to give the customer what (s)he wants, give the leadership what they desire, and provide the people with prosperity, meaning, mastery, and membership from the organization in which they work. Although money is a motivator, money is not the only method of motivation.

Motivation comes from recognition. Motivation comes from success. Motivation comes from satisfaction of a job well done. Motivation comes from a respected leader's gratitude. Motivation comes from customers acknowledging the effort and the work of the employee. Motivation comes from peers and teammates congratulating a team member for good work. Motivation happens when people are permitted to improve the processes and workplace in which they are involved.

Improvement happens when an opportunity for process enhancement is identified, and the root cause of the problem is defined, and the process problem is resolved, and the process is improved for the better. Improvement encompasses; the physical workspace, the actual work process, the work environment, and the work culture. From April 1995 until January 2000, I was involved with a Harvard Research Study on Motivation. The study was led by Dr. Teresa Amabile, assisted by Dr. Steven Kramer (each of them had a research assist), with six field associates, focused on motivation, and titled: The Events And Motivation Study[3] (aka, The TEAM Study). After 5 years of gathering more than 10,000 points of data from hundreds of workers from dozens of work teams, the factors of motivation (and "de-motivation") began to come clear. The results of motivation had a strong correlation and positive affect on performance (i.e., individual performance, team performance, organization performance).

Performance is the result, or output, of processes driven by people delivering value to customers. As Dr. Deming put it, "Every system is perfectly designed to get the results it gets."[4] In order to get better results, a system must be improved. In order to improve the system, the processes within the system must be improved. Hence, if you want better performance, improve the processes in the system. Use the people to solve process problems … they are already being paid for the work, why not use the resource to solve problems to create more quality work, and eliminate "non-quality" work? A simple, sequential order for kicking off a performance improvement initiative; Start with the workplace or work space, proceed to the processes and systems, advance to the work environment, then progress to the work culture (NOTE: This is not "the only" approach that works, but it does move from individual, to group work, to organization, to enterprise).

1. Create a Predictable, Effective, Efficient, Organized Workplace, or Work Space
   The focus on motivational and productive work spaces are a good entry point for establishing Healthcare Affordability. Stable, static, proficient work areas reduce the waste of time, motion, over processing, and waiting. A common resistive response to this recommendation is, "That will cost money to re-design, re-layout and buy new furniture." When often is the case, design, layout and furniture is not required.

   A Lean technique and tool used for this activity is called "5S." It's a simple five step method to organize, standardize, and sustain a work space.
   a. Sort: Sort out what is needed and what is not needed. Determine what is necessary for the work being done. When in doubt— throw it out!
   b. Straighten: Put everything that is needed in an orderly fashion so that things can be accessed easily.

    c. Scrub: Clean —Eliminate the sources for dirt, and the existence of anything that is apt to create "dirt" (duplicates, obsolete). Anything that exhibits "the unnecessary."

    d. Standardize: Make standards so that any abnormality becomes obvious. Emphasize —"The best way to do something."

    e. Sustain: Be Self-disciplined. Sustain the improvements to prevent backsliding. Get management involved to audit the ongoing 5S effort.

This approach creates a cyclical process and a daily event that should be executed at each work location. In many organizations, the last step is the hardest to maintain. The discipline of sustaining the activity usually requires that management and leadership encourage, support and engage in reviewing the results in a regular basis (Suggestion: Weekly)

If done correctly and consistently, 5S touches on a few behavioral and cultural aspects of Affordability. Predictability—one knows what to expect when (s)he enters the work space and is assured as to where things are and how things work. Effective—knowing where things are makes it easier to perform the work, since the organization of the area supports the process or processes in function. Efficient—knowing how things are organized provides efficiencies and proficiencies for the work being done. And perhaps most important, emphasizes the importance of standardization.

After work space organization, new layouts and flow designs can be investigated and implemented as required. The cost savings from workplace organization may be able to fund new work space layout requirements. However, the first step is to convince all involved that better workplace designs can improve Affordability.

2. Improve Procedures, Tasks, Activities, and Operation of the Process
A most important initiative, if not the most important initiative, is process improvement. At the strategic level, the enterprise and the systems of the organization should be mapped and documented (Recommendation: Value Stream Mapping). At the operational level, the value stream maps and system documentation should be documented using process mapping. At the tactical level, the processes should be detailed and supported with the work procedures, tasks, activities, machines, tools, materials, and information necessary for execution of the process.

With the establishment of stability, standardization, and sustainment of the processes, the people can be put to task to improve the processes in which they operate. This requires, measurement, review, identification of opportunities for improvement and activities focused on process problem-solving and improvement. For this to be successful, a few key elements must be present: (1) A regular habit of Problem Finding and Problem Identification, (2) A standard problem-solving process, (3) A work culture that encourages and supports problem-solving as a part of everyday work life, (4) Celebration, Recognition and Reward for Problems Solved and Process Improvement, (5) Management and Leadership involvement and engagement.

3. Establish a Work Environment of Excellence

   A work environment of excellence can be equated to the concept of "World Class" using "World Class Standards." First a definition of World Class must be described. World-Class Standards are qualitative and quantitative measures of a product, service, or organization that are acknowledged, accepted, and admired by customers, stakeholders, professional peers, and competitors alike. They distinguish the product, service, or organization as one of the "best in its class." When we say "acknowledged" we mean the standards are commonly held throughout the world to be true and fair measures. They are "accepted"—understood and embraced—as benchmarks others strive to attain. And they are "admired"—as indicators of an organization that exhibits quality and a value-added reputation, and consistently exceeds customer expectations. A work environment of excellence contains such attributes.

4. Institute a Work Culture of Affordability

   A work culture of Affordability is best described in the foundational text: "Affordability: Integrating Value, Customer and Cost for Continuous Improvement." The work culture of Affordability delivers value that is defined by the customer, generates revenue at a competitive price, and operates profitably through the management and improvement of the expenses required for the resources necessary for value delivery. The integration of value, customer, and cost is the aim of Affordability. The Affordability work culture is customer centered, people driven, process focused, and performance oriented. Key factors in the work culture of Affordability are problem finding and problem-solving. The entire workforce should be put to work finding and fixing problems they encounter in their work environment. This is the heart of the Healthcare Affordability Culture.

The power of problem finding

Over the past 40 years, I've often heard managers say to employees, "Fixing problems … that's not your job. We hire people to fix problems. Just go back to work and do you job." Over time, the employee learns to respond, "That's not my job. Just tell me what to do and I'll do it." Most of us seem to agree, the people best to fix any problem, are those closest to the problem, supported by others who have specific expertise to assist in problem-solving. Even if those resources are in place, problem-solving can't occur, until problem finding, problem prioritization, and a standard problem-solving process is invoked. Step 1: find the Problems, Step 2: Prioritize the Problem, Step 3: Activate the Standard Problem-Solving Process.

It all starts with Problem Finding. Thinking about it from a pragmatic standpoint, who wants to be the one who constantly points out problems? Practically thinking, those who continuously identify problems are often thought of as the problem. It's not natural in the workplace to encourage problem finding, unless the organization emphasizes the identification and solution of problems as a cultural norm. In fact, fixes to problems are often solved by "quick fixes" and band aid mitigations, where firefighting become the normal where crisis responses are rewarded. Healthcare Affordability has an improvement basis of Problem Finding.

Problem Finding is based upon observing conditions and symptoms, digging into the process exhibiting circumstances and identifying root cause scenarios that create those situations, and put in motion a standard problem-solving process for resolving process defects. Teamwork and team based problem-solving is often required to resolve very difficult conditions and circumstances. Basically, the message is, "Put everybody to work solving problems and improving processes."

The impact of problem-solving

For any organization, problem-solving can impact the loyalty and retention of customer, employees, and suppliers. Problem-solving can increase speed, improve quality and lower cost. Problem-solving adds to profitability, demand, and growth. It increases capacity, improves capability, and provides flexibility. For any organization it's often worth millions of dollars, in terms of revenue, cost, and profitability. It creates an environment that re-establishes a "Pride of Workmanship" among the people.

Over time, I've collected a portfolio of examples that illustrate and elucidate the what happens when the people are able to apply problem-solving methods as a regular part of their work. Within Healthcare Affordability, other than the 70 examples I prefaced in Chapter 3, I'd like to offer yet several more examples that make the point of the impact of problem-solving. These examples represent only a portion of the Healthcare Enterprise, and can serve to demonstrate what can be accomplished when the people are put to work solving the problems as a part of their roles and responsibilities.

Examples

Below I'm including some of the examples I've collected of tactical level efforts of improvement over the years. The five that follow span Healthcare Providers, Service Providers, and Device Providers. They encompass the individual aspect, the process dimension, the work environment, and the work culture. When all aspects are included, motivation increases, performance increases, and customer satisfaction increases. This is one facet of the ultimate Healthcare Affordability performance improvement impacting; Customer Satisfaction, Employee Satisfaction, Speed, Quality, and Cost.

Healthcare: nursing supply room project

The problem statement (exactly as written): The current stockrooms on the nursing floor are in the same configuration that they have been in for over 10 years. The current configuration is a group of metal shelving units with generally loose items on them. There are a few sporadic bins and drawers but the overall area is not organized at all.

- Items are not organized
- Items are restocked off sticker charges and can be delayed due to human involvement
- Nurses cannot find what they are looking for even if it is in the room, this leads to...
  - Nurses calling for supplies (Reduces materials management productivity, causes transfers and restock of even one item)
  - Nurses going to another department to get the supplies needed (this causes problems with inventory levels within various departments)
  - Nurses having to come into the Materials Management department after hours for an item (this results in the item being delayed in being transferred to the department's inventory and also not being stickered for charges, also possibly resulting in a lost charge)

Approach (as originally planned): Use a core leadership team of five people (Materials Manager, Two RNs, CFO, Ward Clerk) plus the entire Materials Management Team, to organize all supply areas using Lean and 5S principles as well as changing the ordering system to incorporate a replenish signal to be used for restocking ensuring the smooth flow of materials. Including a method for "new stocking" material that has not been stocked (i.e., new material or better material or replacement material that is "new" to the stockroom). Also including a sustainment procedure for assessing that status, improvement, and maintenance of the implementation(s).

Result

	Before	Goal	After	Meets goal?	Comment
Surveys	Scores were 74% poor	Improve satisfaction	Scores were 70% excellent	Yes	Very positive and enthusiastic feedback
On hand $	$8,303.12	$6,162.18	$4,706.15	Yes	Mostly due to overstock that was reduced
Lost item5 $	$315.10	$252.08	$290.81	No	Close, could be due to item variation
Lost revenue $	$1,391.04	$1,112.83	$1,045.80	Yes	Squeaked in due to variation
Nurse item event	47	11.75	10	Yes	High volume dressings were key
Time to restock	35 min avg	Reduce	22 min avg	Yes	Key was not having to sort through sticker charges

This project focused on process improvement at the tactical level, improved speed, increased quality of service, and reduced cost.

Healthcare: catheter project

This project involved processes and procedures of catheter use. One of the top three wastes in Healthcare occurs from defects. Elimination of defects through process and procedure improvement increases patient quality of life and lowers cost. Where the Opportunity for Improvement exists.

- Foley catheter placement and utilization (duration of device)
- Implementation of Foley removal protocol
- Documentation of Foley insertion and discontinuation to ensure appropriate tracking and data collection.
- Reduction of Catheter-Associated Urinary Tract Infection (CAUTI)

Purpose:
- Implement Foley catheter placement criteria
- Ensure that the Foley removal protocol is consistently implemented
- Improve documentation of Foley insertion and discontinuation in the medical record.

Scope: Emergency Center (EC) and all adult patient care units
- Foley catheter insertion in EC
- Foley catheter utilization for all adult inpatient units
- Implementation of Foley catheter removal protocol—all adult inpatient areas

Steps
- Process Flow Mapping
- Root Cause Identification and Analysis
- Definitions of Interventions
- Facts and Statistics: Device Utilization, Infection Ratios
- Problem-Solving Process: DMAIC
- Team Solution Implementation
- Defect Reduction and Cost Reduction (Baseline: 10/2013–6/2014, Outcomes: 10/2014–6/2015

Number of CAUTIs	Before: 40	After: 20
Additional Cost Associated with CAUTI	Before: $3,380	After: $3,380
Cost of CAUTI	Before: $135,200	After: $67,600

Healthcare: early elective deliveries project

Early Elective Deliveries (EEDs) embody five of the biggest problems in our health system today. EEDs are births scheduled without a medical reason between 37 and 39 completed weeks of pregnancy. They often involve:

- Too Much Unnecessary Care: Overuse and unnecessary care accounts for anywhere from one-third to one-half all Healthcare Costs.
- Avoidable Harm to Patients: EEDs harm women and newborns. Babies born at 37–39 completed weeks gestation are at much higher risk of death.
- Billions of Dollars are being Wasted: The cost of these unnecessary, harmful EEDs was estimated in a study in the American Journal of Obstetrics and Gynecology to be nearly $1 billion per year.
- Perverse Incentives as to How We Pay for Care: The truth about EEDs is that our payment system encourages them. They generate admissions to NICUs, and NICUs are profit centers. Studies suggest that reducing the rate of these deliveries to a reasonable number could eliminate as many as one-half million NICU days, which could lower health costs for the U.S.
- Lack of Transparency: We have far more information available to us to compare and select a new car than we do to choose where to go for lifesaving health care.

Project Lead: Lynne Hall, RN BSN, Georgia Hospital Association, Quality Improvement Specialist.

Project Sponsor: Georgia Hospitals Association

Project Background (NOTE: This was a Statewide Initiative in Georgia)

- CMS (Centers for Medicare and Medicaid Services) and the National Content Developer charged all HENs (Hospital Engagement Networks based on U.S. Standards and Regulations of the Joint Commission Resources) to reduce HACs (Hospital Acquired Conditions) by 40%
- Adding reducing readmissions by 20%
- Adding reducing EEDs by 40%
- CMS added EED as a first focus project—meaning results needed to be fulfilled by August 2012

Project Vision: Decrease EEDs by 40% by August 2012

Project Mission: To implement "Hard Stops" (a process that empowers nurses and schedulers to say no to physicians who want to deliver mothers prior to 39 weeks without medical necessity) in order to decrease EEDs for better health of Mother and Baby.

The Approach: Lean Six Sigma

- Hoshin Kanri for Strategic Planning
- Value Stream Mapping
- Process Flow Diagramming and Process Mapping
- Problem-Solving Using IISE DMAIIC (NOTE: The IISE uses a second "I" for Implement)

The Implementation Highlights

- Telnets/Webinars and one-on-one calls with hospitals were held; subjects included:
  - Physician engagement
  - "Hard Stops" policies and procedures
  - Risks to moms and babies prior to 39 weeks gestation
  - Barriers to implementing "Hard Stops"
  - Tools available from March of Dimes to help implement "hard stops"
  - Effects on newborns prior to 39 weeks
- These techniques along with attendance have proven to be successful methods in which to promote motivation and encouragement
- In addition to flow-charting their own process, hospitals were also given options to:

- View all recorded sessions at a later date should a hospital want to view it again or if session was missed
- Use the resource page which included examples of Policies and procedures of best practice hospitals
- Access and use the March of Dimes toolkit
- Flow charts were encouraged to be shared and reviewed
- Several hospitals shared their Best Practices in reducing EED's
  - Atlanta Medical Center
  - Emory University Midtown
  - Piedmont Henry Medical Center
  - Liberty Regional Medical Center

EEDs from Baseline to Result

- 2009—65%
- 2010—35.3%
- March 2012 Baseline—8.77%
- August 2012—3.67%
- YTD 2012—5.90%
- That's a *58% decrease* in EEDs from our March baseline and
- A 90% decrease in EEDs from the 2010 baseline

More Good News

- There are 83 birthing hospitals in Georgia
- 58 (70%) of those hospitals turned in data
- 19 (31%) of the 58 hospitals were already at a 0% EED rate
- Of the 39 hospitals needing improvement about ½ showed a decrease in EEDs
- 3 of those hospitals went from a 14% or higher EED rate to a 0% rate sustained for at least 3 months!!
  - One hospital went from a 30% EED rate down to 0% and has sustained the rate for 4 months

Financial Impact

- According to Managed Care Magazine it costs around $41,000 for a late preterm NICU visit
- The incidents went from 148 incidents in March 2012 to just 33 in August 2012
- That's a decrease of 115 incidents
- Conservatively estimating ¼ of those babies did NOT go to the NICU, and we saved Georgia Healthcare $1,178,750.00 ... OVER 1 MILLION Dollars!! (The actual amount is likely much more).

Outcomes

- Freed Resources
- Better Care
- Better Quality of Life
- Lower Cost

Lessons Learned

- Important to work as a team
- Get physician buy-in and have a physician champion
- Empower your schedulers and nurses
- Have a peer review for non-medically necessary EED
- Educate patients early starting at first visit
- Collaborate with others even outside your hospital:
  - Share best practices
  - Share forms
  - IHI
  - Other Stakeholders such as The March of Dimes
- Use data to sustain the gain
- Present data to administration and physicians
- Build on existing relationships
- Celebrate your success!!!
- The partnerships developed through this project have and will lead to further projects related to mother/baby care
  - Infant Mortality Coalition
  - Safe Sleep Campaign support
  - Baby Friendly Hospitals

Service provider: resource scheduling project

A clear challenge in every Healthcare facility is scheduling. This project involved a partnership effort between a Service Provider and a Healthcare Institution ("The Customer").

The Customer

- A Top Ten Interventional Cardiology Department
- ~11,000 cases per year
- 10 Cath Labs, 6 EP Labs (ElectroPhysiology Lab), 2 Hybrid Labs
- Large Staff: 80 RNs/CVTs/RTs, 30 Physicians
- Hours of Operation: 24 × 7 (2 On Call Teams Always Available)
- High Volume TAVR (Transcatheter Aortic Valve Replacement) Center

The Problem

- The cath labs complex scheduling (weekly, vacation) process is leading to disengagement and dissatisfaction resulting in excess absence calls and overtime.
- The current staffing schedule does not use their staff to full capacity.

Project Goals

- Increase staff satisfaction with an improved and transparent scheduling process.
- Reduce overtime spend by 10%
- Add five more daytime case slots per day

Methodology

- Initiate Engagement
- Assess: Current State, Strengths, Weaknesses
- Develop Solutions
- Implement Solutions
- Monitor Solutions

Project Charter

- Project Sponsor
- Team Leader
- Physician Lead
- Service Provider Team members
- Customer Team Members
- Timeframe: Start Date—End Date
- Problem Statement
- Scope
- Goals
- Business Benefits
- Deliverables

Main Tools Utilized

- Workflow
- DMAIC
- SIPOC
- Root Cause Analysis: "Cause and Effect Diagram"
- Effort vs. Impact Grid

Results

- Predictability up 500%
- Vacation View from ~3 weeks to 12 months
- Reduced no. of shift options by 2
- Schedule creation time down by 50%
- Staff engagement up from 43% to 87%
- 43% reduction in overtime spend from $16K/month to almost $8K/month (almost $100K annually)
- An additional ten more daytime case slots per day were added

Device provider: multiple projects

Several years ago, a Medical Device and Supply Provider embarked on an Affordability journey to increase process speed for greater capacity and responsiveness to demand, improve quality and eliminate rework, and lower cost and decrease expense. The effort was primarily based on process improvement projects and kaizen events (aka Rapid Improvement Events). After receiving Lean training, participants were required to complete a project or kaizen at the tactical value stream level before they received final certification. There were more than two dozen projects executed. Below are briefs of just a sample of their results:

Event: Electronics Room Process Improvements

Description: Because of the increase in demand, with a Takt Time of 1.61 hours, the electronics room could not keep up with customer pull and orders, with the result being overtime expense.

Goals: Map the value stream, identify waste, improve the design and layout, and implement the new process to meet and exceed customer demand. No numeric values were projected.

Result: After mapping the Value Stream and Process Mapping, executing 5S events, instituting several process improvements, and designing and implementing a new layout for flow, the following results were realized:

- Cycle Time: From 2.0 to 1.25 hours
- Lead Time: From 330 to 160 minutes (51% improvement)
- Material Distance Traveled: From 196 to 110 ft (44% improvement)
- Throughput: From 3.6 to 5.8/day (61% improvement)
- Units per Day: From 7 to 11 units (61% improvement)
- Annual Cost Savings: $10,586.00

Event: Print on Pouch

Description: The use of labels on pouches results in a great number of defects and scrap 90% of the products manufactured are packaged in pouches. The printing and application of the labels results in higher labor and material costs which can increase the sale cost of product as much as 25%.

Goals: In general, the overall objective of this project is to reduce cost and eliminate waste from the current procedure through the implementation of the pouch printer.

Result: After implementation process speed increased, quality improved and cost decreased:

- Speed: From 180 to 225 pc.s/hour
- Quality: Elimination of Label Defects due to direct printing on pouch.
- Cost reduction of 25% on labor
- A four month payback period was realized after implementation
- Eliminated the use of a physical label, eliminated label transportation, and eradicated label scrap
- Cost saving of $65,956.00 per year

Event: Incoming Inspection

Description: Every component receives some type of inspection by QA personnel.

Goals: Assess and evaluate the process. Determine requirements and classify each for either inspection or audit. Eliminate waste from the process.

Result: After following the four step improvement method: Map the Current State, Identify Waste, Map the Future State, and Design and Implement Solutions, the following outcomes were realized:

- Process Steps: From 21 to 10 (DTS 52%), From 21 to 19 (INS 9.5%)
- Time: From 749 to 45 minutes (DTS 94%), From 749 to 528 minutes (INS 29.5%)
- Lead Time: From 1.5 to 1.1 days (26.7%)
- Walking: From 1351 to 1140 Steps (16%)
- Annual Cost Savings: $229,770.00

Event: Solutions: Sterilant Line

Description: Although quite efficient, this team focused on instituting continuous improvement on the process and in this work area.

Goals: Initial Goals Included—Reduce lot changeover time 20%, flushing 20%, Filler set up time reduction 15%, Establish standard work for setup, shift changes, chemical/bottle changeovers, 5S, five safety improvements

Result: Several problems and opportunities were identified for improvement, as well as numerous safety improvement options. The following was the consequence of team improvement actions and activities:

- Productivity (Goal—Reduce Setup time by 20%)
  - Process Stage 1: From 41.5 to 16 minutes (60% decrease)
  - Process Stage 2: From 79 to 39 minutes (50% decrease)
- Movement/Steps
  - Process Stage 1: From 1197 to 200 ft (83% decrease)
  - Process Stage 2: From 2860 to 1372 ft (52% decrease)
  - NOTE: A distance pf 122 miles less per year of walking
- Implement 5S: Before—None, After—100%
- Standard Work: Before—Nothing Documented, After—100% (Better Flow and Greater Output)
- Safety: Goal—five improvements, Result—six improvements

Event: Mold Preventive maintenance

Description: Because of the mandate of sterility for these products due to medical use, mold prevention is a mandatory requirement.

Goals: (1) Reduce our mold preventive maintenance turnaround time. (2) Implement a dedicated mold preventive maintenance area. (3) Develop standard work instructions for mold preventive maintenance.

Result: After several kaizen events, the team realized the following improvement:

- Average Turnaround Time (hours): From 32 to 4 per mold (−28 hours, −88%)
- Average Turnaround Time (in minutes): From 1920 to 230 per mold
- Process Steps: From 11 to 5 (−6, −45%)
- Walking Distance (ft): From 490 to 74 (−416, −85%)
- Savings
  - Savings Direct Labor: From $240,000 t0 $40,000
  - Tool Maker freed for other Tooling: $240,000
  - Transportation Reduction: $15,000
  - Downtime Savings: 125,000$
  - Total: $575,000.00

Event: Lot Release Process

Description: After mapping the processes for sterile and non-sterile products, identifying the waste, assessing the current conditions, developing future state flow maps and spaghetti diagrams, and developing improvement plans, several events were held to implement the improvements identified by the team.

Goals: Reduce walking distance during lot release process. Improve product flow, especially for kit staging.

Result:

- Lot Release Time—Sterile: 51.02 → 50.2 hours/week (−0.82, −1.6%)
- Lot Release Time—Non Sterile (boxed): 43.94 → 3.25 hours/week (−40.69 hours/week, −92.6%)
- Lot Release Time—Non Sterile (loose): 46.09 → 0.00 hours/week (−46.09 hours/week, −100%)
- Product Traveled Distance: 37,919 → 33,304 ft/week (4,615 ft/week, −12.2%)
- Walking Distance: 29,180 → 6,746 ft/week (−22,434 ft/week, −76.9%)
- Annual Walking Distance per Year Reduced: 1,121,700 ft/year or 22.44 minute/year
- Cost/Savings: $1,475.18/week → $1,315.65/week (−159.53, −10.8%)
- Annual Savings: $8,295.56

The examples described above are only a subset of all the activities and actions taken. Although this effort is ongoing, the financial savings alone (including the above examples and those not included here) amounted to over $5 Million! This reduction in cost frees up money to be reinvested, frees up time to be reallocated, frees up margin to create even more competitive pricing.

By focusing on process improvement, considering both qualitative measures (Customers, People, Suppliers) and quantitative measures (time, quality, cost), it is proven, by example and result, "Faster, Better, more Affordable Healthcare" can be achieved.

Reviewing performance and results

A key factor of success in all of these efforts includes performance reviews on a regular basis, and celebration and recognition of results. The results of every one of these efforts come from the people solving process problems. The work that is being done is by way of a process or processes in the systems of the enterprise. The results and outcomes just provide a snapshot of performance, where the real gains are made within the process improvement effort. Leadership (at the strategic level) and management (at the operational level) must put a structure in place to regularly review the status of the people, the process(es), and the performance, and recognize and celebrate wins and successes. Although this is being done at the tactical level, it must link to the operational level, and to the strategic level to assure alignment. A structure should be developed that measures at the value stream level with the same metrics being measured at the strategic level, and all are linked through the operational level.

There are key common themes at all three levels of the organization that involves customers, the business, the processes, and the people. At the tactical level, this can be accomplished by creating "performance boards," sometimes called "huddle boards," where a team or unit can gather to check status, review information, recognize and celebrate victories, understand the current condition of problem-solving endeavors and activities, appraise results measures, evaluate resources requirements, and address needs for the day's work.

The customer, to whom the value is being delivered, is fully integrated in the process. The process is the work of the people, and the people's business is the integration of the expenses and resources required, and the price and revenue being generated. There are three major points of review necessary; People (the team members, their credentials and certification, the team resources required, team information), Process (the process itself, the procedures, tasks, machines, tools, material, supplies, information), Performance (Qualitative: Customers, People, Suppliers and Qualitative: Time, Quality, Cost).

The team should hold a performance review daily (During a huddle lasting about 15 minutes). Management should review the team's performance weekly. Top leadership should be involved in a review monthly. In one good example with which I'm familiar, the team's huddle is at 8:15, management's huddle is at 8:45, senior management's huddle is at 9:15, and top leadership's huddle is at 9:45 every day. With this design, any serious or critical issue occurring at the value stream level can be communicated up to leadership before 10:00 each day. This particular rhythm is designed to address, in 15 minutes, customer conditions, resource need, identification of opportunities for improvement, improvement activities, recognition and celebration of accomplishments, and a special component called "Good Catches" (i.e., when someone catches a defect or defective condition, and either communicates the situation to management, or implements a solution, they get recognized for a "Good Catch").

At the tactical level, success in terms of value, customer, and cost must be fully integrated within the organization. Of course, Affordability is based on the integration of value, customer, and cost for continuous improvement. To achieve such ends, the means of Affordability must be reviewed, assessed, addressed, and realized. In some rare cases, self-directed work teams have been implemented by leadership. With the establishment of purpose, direction, alignment, motivation, communication, and execution, a strong linkage and connection can be established, and the value stream can be operated at a profitable level, delivering value to the customer per conformance to requirements and compliance with standards.

Here's a List of 10 Best Practices at the Tactical and People Level for Healthcare Affordability:

1. People solve the problems in the process and the workplace.
2. People are enabled to act and improve the workplace.
3. People are part of the solution, not the problem.
4. People are provided with the tools, methods, and resources to solve problems.
5. People are accountable and responsible for Performance (leadership, management, all people).
6. People are recognized and rewarded for accomplishment and achievement.
7. People have a sense of: Meaning, Mastery, Membership (Dr. Rosabeth Kanter) for what they do.
8. People are applied for the skills they have, their own capabilities, and the passion they possess.
9. People are either, Value Add or Value Added Support, for the Delivery of Customer Value.
10. People are compensated at a competitive level and above.

References

1. The W. Edwards Deming Institute Blog, 2018, W. Edwards Deming on People, https://blog.deming.org/w-edwards-deming-quotes/large-list-of-quotes-by-w-edwards-deming/.
2. Margaret Wheatley's Ten Principles for Creating Healthy Communities, 2010, https://markholmgren.com/2010/03/29/10-keys-to-healthy-community-change-%E2%80%93-margaret-wheatley/
3. Teresa M. Amabile, Sigal G. Barsade, Jennifer S. Mueller, Barry M. Staw, Affect and Creativity at Work, *Administrative Science Quarterly*, 50 (2005): 367–403.
4. The W. Edwards Deming Institute Blog, 2018, http://quotes.deming.org/authors/W._Edwards_Deming/quote/10141.

chapter seven

First and foremost, assess the situation

> To know where to go, determine where you are,
> discover where to go, then plan your journey.

Paul Walter Odomirok

It's not surprising that many organizations relive the journey they're taking many times. Once, while engaged in a Lean program, I discovered the organization tried Lean in 2000, 2002, and 2004, only to fail and start all over again. It seemed that the lessons learned at each attempt were not considered when the next attempt was made. The eventual success in the program was in part due to overcoming and mitigating the pitfalls and roadblocks of the past three tries. Without the knowledge of the past and the current condition, the fourth try would have likely failed. And in keeping with history, the George Santayana saying, "Those who cannot remember the past are condemned to repeat it," applies to Affordability as well.

It's not only the past but the current state too. Within the current state reside the clues, signs, and opportunities pointing the way to success. By solving the problems in the enterprise systems, and instituting better ways to deliver value, the successful results of Affordability can be fulfilled. Although it seems so straightforward and simple, the endeavor is complex and challenging. The processes really don't care if they change, the people do. So to fully assess the current condition, include both the processes and the people.

First of all, research and understand as much as possible about the strategy, systems, and structure of the organization. In Chapter 6 of *Affordability: Integrating Value, Customer, and Cost for Continuous Improvement*, a number of tools and techniques are provided for this activity. Determine the maturity of information and data, and build the profile according to the need defined. Taking time to understand will be paid back during implementation when hidden barriers appear. Be sure to understand the current:

- Purpose: Why we do what we do, and why we exist.
- Vision: Where we are heading.
- Values: Our organization's norms, standards, and principles.
- Mission: How we're going to get there.

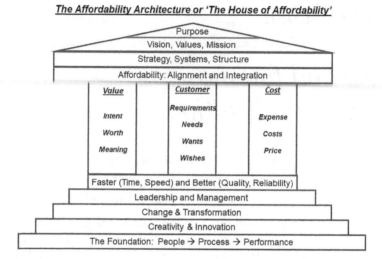

The Affordability Architecture or 'The House of Affordability'

Figure 7.1 The architecture of affordability or house of affordability.

With this understanding, top leadership can be engaged to create the new direction going forward by developing the new purpose, vision, values, and mission. The "House of Affordability" may be helpful in framing thinking (see Figure 7.1).

Customer, value, and value stream

Since Healthcare Affordability is Customer focused and Value driven, a comprehensive knowledge base of this information serves to lay the foundation for an Affordability design. Answers to questions, like those below, are only the beginning of a complete assessment:

Customer
The customer is the one who is willing to pay for what the organization offers. In terms of a patient, it is the one who receives the care that is provided. In terms of providers to Healthcare, it is the product or service required to deliver care to the patient. Some questions to answer include:
- Who is "the customer"? Is the customer, or are the customers, clearly defined and understood?
- What are the different customer types and categories?
- Are we customer focused or corporate focused?
- What are the markets and the market segments by the value we deliver?
- Who is the competition and how do they deliver care?

Value

The Value is the product or service the organization provides to satisfy the requirements of the customer.

- Is the value defined by the customer or the organization?
- What is the Value of the organization?
- What is the value we deliver?
- Is the value delivered by way of a product, or service, or both?
- Is the Value, Non-Value, and Value Added Support concepts known and understood?

Value stream

The Value Stream is the method and manner the Value is delivered.

- Exactly how is the value delivered?
- What are the Value Streams and value delivery methods?
- Is each Value Stream Mapped?
- Do stabilized, standardized, and sustained processes exist that support the Value Stream?
- Do policies and procedures exist for all work processes mapped and documented?

Here are more general questions for the Affordability Assessment:

- What do we do that the customers want?
- What are we not doing that the customers want?
- What are the customers' requirements?
- What are the requirements beyond the customers' scope?
- How do we deliver our value today?
- How will we deliver value in the future?
- How will we create more value for delivery in the future?
- What resources do we have, and use, for delivering value?
- What resources do we need to create and deliver the value?
- What are the expenses for resources, value creation, value delivery?
- What is the price of the value delivered?
- What is the revenue flow, profile, and perspective from customers purchasing the value?
- What is balanced in terms of Value, Customer, and Cost?
- What is unbalanced?
- What is the customer demand?
- What is the potential customer demand?
- What are the opportunities to improve upon value delivery per customer requirements?
- What are the opportunities to reduce expense and better optimize resource provisioning?
- What are the opportunities to reduce price and increase revenue?

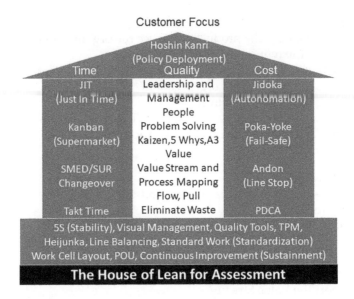

Figure 7.2 A framework for assessment.

In addition to "The House of Affordability" above (Figure 7.1), "The House of Lean for Assessment" (see Figure 7.2) may be beneficial tool to use for answering questions, identifying gaps, and developing Lean Thinking.

Strategy

Having been a corporate strategic planner, I realized that companies often base their strategic plans on their financial perspective. Taking such an approach provides strategies from lagging indicators, and subjugates leading indicators. It defines the future primarily on what's occurred in the past, ignoring future possibilities, opportunities, and potentials. Although this approach tends to provide a direction of uniformity and regularity, it does not institute change and progress. A good framework and foundation for an Affordability strategy are the 14 Lean Principles

1. Base your management decisions on a long-term philosophy, even at the expense of short-term financial goals.
2. Create a continuous process flow to bring problems to the surface.
3. Use "pull" systems to avoid overproduction.
4. Level out the workload (work like the tortoise, not the hare).
5. Build a culture of stopping to fix problems, to get quality right the first time.

6. Standardized tasks and processes are the foundation for continuous improvement and employee empowerment.
7. Use visual control so no problems are hidden.
8. Use only reliable, thoroughly tested technology that serves your people and process.
9. Grow leaders who thoroughly understand the work, live the philosophy, and teach it to others.
10. Develop exceptional people and teams who follow your company's philosophy.
11. Respect your extended network of partners and suppliers by challenging them and helping them improve.
12. Go and see for yourself to thoroughly understand the situation.
13. Make decisions slowly by consensus, thoroughly considering all options; implement decisions rapidly.
14. Become a learning organization through relentless reflection and continuous improvement.

These principles are beyond customer, value, and value steam. They encourage leadership to focus on the long-term, flow, balance, problem-solving, standardization, continuous improvement, visuals, technology, leaders, suppliers, observation, decision-making, learning, and growth.

Since a Healthcare Affordability Strategy may be built on a time horizon of 3–5 years or even more, there are four phases to consider: (1) The initiation phase focused on kicking off the strategy, creating momentum, implementing Quick Wins, and establishing proof of concept; (2) The second should continue progress leveraging gains and successes to create more change and improvement; (3) The third phase should emphasize the institution of successful changes, standardization of the organization's activities, and celebration of successful performance improvements; (4) Finally, the last phase is the culmination of the current strategic plan, the designing and planning for the next strategic program, and the linkage between the current strategy and the new strategy.

Assessing the current strategy for designing the new strategy includes discovery of: the current direction, the alignment of resources, the motivation of the people, the message and communication of the message, and of course, the design and plan for execution. On the human behavior side, especially for leadership, the following attributes should be sought: (1) Does leadership "follow the way" and behave in the prescribed manner?; (2) Does leadership inspire a shared vision for action?; (3) Does leadership challenge the process and the way of doing things, instead of challenging and blaming people?; (4) Does leadership enable others to act, solve problems, and improve the system?; (5) Does leadership encourage the heart, foster teamwork and collaboration, and persuade all to get involved?.

Systems

For any element or business within the Healthcare Affordability Enterprise, the enterprise level systems that deliver value, support value delivery, and manage the organization, serve as the overall aims for improvement within the pursuit of Affordability. The processes that are within each of these systems are the targets for problem-solving and improvement. Whether they are diagrammed and documented, or not, standardized or sustained, or not, a thorough understanding must be established for knowledge and awareness to know what to do next. A simple approach to begin this assessment would be to determine system and process maturity:

1. Stabilized? Are the systems and processes chaotic and inconsistent?
2. Standardized? Are the systems and processes standardized in terms of documented and stable?
3. Sustained? Are the systems and processes sustained and maintained in an effective manner?

The results of such a study would activate the next step and suggest the next response:

- If not stabilized, then stabilize: Document, diagram, map, and focus on removing chaos.
- If stabilized, but not standardized: Work toward consistency, eliminate waste, and increase flow.
- If stabilized and standardized but not sustained: Increase performance, reduce variation, sustain, and maintain an effort of continuous improvement.

As for the human factor, or people systems, there must be a clarity of what is often referred to as staffing, style, and skills. The specific resources, alignment, competency, capability, capacity, dynamics, principles, values, purpose, meaning, mastery, and membership factors must be clear. For designing and planning purposes the attributes, roles, and responsibilities for all of the at the top leadership and senior management, management, value stream resources, and value stream support resources attributes should be clearly understood and described. These so-called people systems are sometimes looked at as secondary to the business, but in reality, they are a key component of the business. Although anyone is replaceable, whole systems of people are not.

Structure

With full knowledge of the systems and personnel, the structure of how processes and people are interrelated provides the fully integrated perspective of how the organization currently operates. The assessment of structure should contain the formal and informal dynamics within the organization. Organization charts, documented roles, and responsibilities, even work instructions provide the formal dimension. However, discovery of the informal dynamics might take time and extensive involvement and observation to truly comprehend what the people really do, and how the processes really work.

I can remember an incident during a project with a device manufacturer where I asked the question, "Why do you do it that way, when the process documentation calls for a different method?" The response was, "This is a better way to do things. As for the documentation, that's necessary for our ISO Certification." So for processes and people, both dimensions together illustrate a better image of the true structure and how resources operate.

In addition to strategy, systems, and structure, a second dimensional perspective is beneficial. This angle provides more depth and breadth for the collection of artifacts, articles, and information of the assessment. At the strategic level, top leadership and senior management set the strategy, direction, and manage the business. At the operational level, the management focuses on daily activity and overall value delivery. At the tactical level, value is produced and delivered to the customer. Too often, the various divisions and departments at each level, and for each function of every individual, operate as "independent silos," not integrated, not interrelated, and even sometimes competitive. The "Silos of Operation Structure" is detrimental to Affordability.

Strategic level

The strategic level is where top leadership and senior management reside. Their primary role is to move the organization forward through growth and success. This requires change and transformation over time to continue to deliver value to the customer and provide new requirements as they emerge and appear, as well as maintain competitiveness within the marketplace. Some successful managers find themselves promoted to this level and are unprepared to take on the new role required by this echelon.

Today, it is often, that organizations suffer from too much management, and not enough leadership. At this level, leadership is the key to success. Setting direction, aligning resources, motivating people, communicating the message, executing the plan, modeling the way, inspiring a shared vision, challenging the process, enabling others to act, and encouraging the heart are all required actions and behaviors at this level. The outcome of an in-depth assessment would reveal the overall status, and the individual status of every strategic level participant.

Operational level

This is the level where management takes place. Managers at this level provide stability, status quo, and constancy of operation. There is a natural conflict between this level and the strategic level in terms of the requirement to change (strategically) versus the necessity of steadiness (operationally). Managers are responsible to deliver value and accumulate revenue for sustenance. Managers maintain and control resources and expenses. Managers own processes and procedures. Managers manage, while leaders lead (However, it is important to note that it is incumbent that managers lead the people who drive the processes).

The management of operations (Value Delivery, Value Added Support, Business Support, etc.) permits the company to provide and continue to exist. It permits Care Providers to Care for Patients. It permits Medicine and Pharmaceutical Providers, Machine and Device Providers, Service and Supply Providers, and Insurance and Payment Providers to Deliver Value to the Care Providers for Care Delivery. With full integration across the Healthcare Enterprise, Healthcare Affordability can be accomplished.

Tactical level

At this level, are the people that drive the processes that ultimately deliver value. It must be assessed from both a qualitative and quantitative perspective. That is, qualitative in terms of the human factor of satisfaction, with extended consideration to customers (internal and external), and suppliers (internal and external), including personal meaning (the meaning of work being done), mastery (the ability to pursue, grow, and accomplish some level of mastery of the function and the work), and membership (teamwork, collaboration, co-ownership, etc.). The realization of the work being done is through processes; therefore, quantitative factors should be investigated (capability, capacity, time, quality, cost, etc.).

The relationship with leadership and management serves to provide some insight into the readiness for change and preparedness for transformation. Any Affordability pursuit requires change and transformation of the organization. During the initial steps, or even during an advanced part of the journey, change, and transformation is required. The peoples' readiness for change is a critical factor for success.

Assess step by step

Below is a step-by-step sequence for assessing the current state and readiness, as well as serving to discovery opportunities and possibilities for the design and plan. (Please refer to Appendix A for this section.) Although it may take time, it is recommended that every emerging initiative be researched and understood before any undertaking commences.

1. Organize a team to research and understand the current state of the organization.
2. The team should continue this pursuit until they come to consensus that enough understanding has been accumulated. This includes recording gaps and needs discovered while researching. Warning: Avoid paralysis by analysis—The 90% is enough rule may serve to keep the effort from experiencing stagnation.
3. Assess the strategic level, leaders, and the operational level, managers, for continuity of strategy, systems, and structure, as well as propensity for change and transformation, and readiness for an Affordability initiative. This also includes additional gaps and needs discovered. Again, use the "90% Rule" if it's necessary to keep the team moving.
4. Conduct a deep dive into the people dimension. This includes all strategic, operational, and tactical aspects. It also includes; strategy, systems/processes, structure, skills, style, staffing, and commonly shared perceptions of reality.
5. Document the existing gaps and needs, then move on when the team comes to consensus in terms of completion.
6. Finally, investigate all resources, tools, methods, means and standards for people, process, and performance.
7. Move on to Design when complete in accordance with team consensus.

Please note, there are times when in design, additional research, understanding, and assessment is necessary. However, if done correctly, more than 90% of the necessary information should not be available for the team to design, plan, and act.

chapter eight

Design the solution and plan for success

For a solution and plan for success, Healthcare Affordability is truly achieved with the people of the organization solving process and performance problems and instituting Improvements as part of the ongoing and regular behavior in the culture. Suggested Pre-Reading; *Affordability: Integrating Value Customer and Cost for Continuous Improvement*, Chapter 9 "People: The Human Factor."

The role of the change and transformation champion is to; "Get the 'fisher-people' to be able to fish." If you're one of the "fisher-people," fish, catch the solution, and implement it. If you're the change champion, it's up to you to train, develop, coach and mentor the "fisher-people" in the organization. Too often, consultants are hired to "fix the problem," or other people are brought in from outside the process to make improvements, however, the best approach is to (as Dr. W. Edwards Deming put it in his "Point 14 of the 14 Principles"[1]), "put everybody in the company to work accomplishing the transformation." Using outside resources may be necessary during the initial phases to address situations "alien" to the institution, and from time to time with new concepts, but the main goal for the organization is for every internal participant to be improving the workplace for the better.

The best program solution design I've witnessed, involved with the design and deployment phase, occurred in 2007. Although it was not healthcare, the program met its goals, and accomplished the transformation intended. Since that time, I've been able to use the same design and approach with Healthcare Enterprise institutions. The solution contains a five key components critical to success: (1) purpose and direction, (2) leadership, (3) people, (4) processes and resources, (5) plan and program.

The purpose and direction should be clearly communicated and understood by the leadership, management, and every one of the people in the organization. The purpose can be a response to a crisis or reaction to a gap or need for improvement for increasing market demand or contending with a formidable competitor. The direction should include the reason and intent, an in addition, the vision, mission, goals, and objectives for the program and the activities within the plan. It is necessary to set the direction, align the resources, motivate the people, communicate the message, and execute the plan for the sake of clarity and specificity.

Leadership, including top leaders, senior managers, and all management personnel, should have a role and responsibility as designated in the solution design. Leadership should be engaged, involved, and participative in all dimensions and aspects of the program. In fact, top leadership should be an integral part and play active roles during the initial activities. It is best when the top leader participates in the first set of activities with the rest of the levels of the organization.

Every person should have a role to play within the plan and program. The first round of activities should include those individuals who demonstrate and display favor and advocacy with the principles of the change and transformation of the new program. This strategy helps to establish early and quick wins, and stimulates momentum for the program. Not including potential resisters in the early phases avoids delays and roadblocks in terms of progress and advancement. As time goes on, and successes accumulate, resistance diminishes, and progress occurs.

Processes and resources for change should be available, developed within the workforce, and supported by all levels of management. Processes for problem-solving, communication, improvement, and change should be trained and developed within the organization. Resources to support the transformation to institute process improvement should be known and available for implementation. This can be accomplished by taking 12–18 people at a time, 3 days in a week, covering the foundation content of the transformation, the change tools and techniques, and chartering a project for improvement. A follow-up 3-day session should be scheduled for the same group a month later to review status, recognize accomplishments, and celebrate victories. Top leadership should be included in the first group to establish a firm sense of commitment to the program. A typical schedule for this approach could occur as shown in Table 8.1.

Using this format and rhythm, a range of 72–108 people can be trained, developed, and achieve accomplishment with project outcomes that include increased process speed, improved quality, and reduced cost. With the inclusion of leadership, and every level of the organization, the entire organization can witness change at all levels.

Last, but certainly not least, the plan and program is crucial for the probability of success. In many cases, a 5 year horizon is appropriate. The first year should be dedicated to initiating the program and creating momentum (see Appendix B). The outset for the first few months should be dedicated toward the deployment of the purpose, vision, mission, goals and objectives,

Table 8.1 Transformation to Institute Process Improvement

Group	Size	Date	Outcome
1	12–18	January 3-Day	Basics, foundation, framework, project charter
1	Same	February 3-Day	Advanced development, project completion/status
2	12–18	March 3-Day	Basics, foundation, framework, project charter
2	Same	April 3-Day	Advanced development, project completion/status
3	12–18	May 3-Day	Basics, foundation, framework, project charter
3	Same	June 3-Day	Advanced development, project completion/status
4	12–18	July 3-Day	Basics, foundation, framework, project charter
4	Same	August 3-Day	Advanced development, project completion/status
5	12–18	September 3-Day	Basics, foundation, framework, project charter
5	Same	October 3-Day	Advanced development, project completion/status
6	12–18	November 3-Day	Basics, foundation, framework, project charter
6	Same	December 3-Day	Advanced development, project completion/status

and creating initial quick wins designed to stimulate impetus for success and achievement. Year Two is focused on leveraging Year One improvement and taking the effort to the next level. Years three and four should be adjusted and synchronized with the necessary improvement and modifications required from years one and two. The final year of the five year plan should finalize the first wave, setting the stage for wave two, and the next strategic plan.

All five components should be in place, at a defined and comprehensive level, before beginning the program. Typically, the preparation and preplanning can take a month or two, and even up to a year or more. Key personnel for inclusion are; top leadership, a program champion, select personnel, and often a "black hat" (someone from the outside that is knowledgeable, proven capable, and trusted.). It has been published several times over the past 20 years that transformation efforts fail 70% of the time. Many of these failures are due to lack of proper planning, while others can be attributed to resistance and lack of the ability to achieve cultural change and transformation.

In addition to deployment of the improvement program, attention should be paid to performance measurement, reporting, communication, status review, and identification of further improvement. Eventually, sustainment of continuous improvement will be a challenge. Continual advancement of enhancements tends to wane after some period of time. It could be 12 months into the program, 18 months into the program or even 24 or more months into the program. A portion of a good solution design should contain a plan for sustainment and revitalization of the effort.

The three core elements of this approach and the solution design show cased here are; people, process, performance. Often organizations will also plan for, and address; communication and messaging, reward and recognition, celebration of success, and a detailed execution project plan communicated to all employees, and even to select suppliers and customers.

Of course, solution designs can come in all "sizes," "shapes" and "colors." Some organizations believe the solution comes through training. I partially agree. I fully support training *and* development. It's the development of the people where the capability is realized. Training transfers knowledge, but development enables people to act. In addition, some component of application is beneficial. An activity, a kaizen, or a project goes a long way to shape and hone behavior. Especially if the actions are focused on process problem-solving or workplace improvement.

For example, in one solution design, a group of 18 employees, representing three areas received 3 days of training and 2 days of development. That first week was followed by 3 weeks' time in which process and workplace improvement activities were executed. A month later, the group came back together for review of results and some advanced training followed by recognition and celebration led by leaders on the last day. This was carried throughout the organization for 3 years and more than 300 people served as points of deployment for the improvement initiatives. If each coached and mentored 10 others, 3,000 people were exposed, and over 50 work areas benefited from the improvement efforts. Process speed increased, quality improved and cost went down ... Affordability!

In yet another design, within an organization of four distinct sites and locations (i.e., Tulsa, Oklahoma; Longview, Texas; Little Rock, Arkansas; Brampton, Canada) key management personnel were trained and developed using a custom designed curriculum and plan for deployment. Each manager, a process owner, championed efforts in their area for process and performance improvement. Teams of people in each group, led by the manager, were put to task to find, fix, improve their own processes. As a result, dozens were developed, hundreds were involved in the initial implementation, and thousands were exposed within the first year, or phase 1.

Regardless of the design, the 5 elements of success were present; Direction (Purpose, Vision, Mission, Goals), Leadership (Involvement and Engagement), People (Participation and Inclusion), Resources and

Processes (To accomplish the tasks at hand), Design and Plan. A key factor of success included the tools and tooling for implementation, that prescribed resources and processes for use and utilization.

Process improvement tooling

The first step is to focus on eliminating waste and create processes that are stable, standardized, and sustainable. This is often accomplished using basic quality tools and Lean methods. Once the process is stable, standardized, and predictable, the application of 6σ methods could be used for further improving process performance and reducing variation. If a process or system, or the processes of a system, are stable, standardized, sustainable, and predictable, the 6σ methods should be applied immediately to further improve process performance, reduce variation, and advance standardization. In addition to the tools and tooling available from numerous sources (i.e., ASQ, IISE, TQM, Lean, Six Sigma, etc.), I've focused on a few key tools and toolsets I've discovered to be most helpful.

System tooling—The 7 Flows

One of my favorite tools in designing and planning a Healthcare Affordability Enterprise Solution is system toolset known as; The 7 Flows. I've learned from my exposure to Lean, going back to 1989, even before they called it "Lean" (known then as "J.I.T." or Just In Time Manufacturing), that after one realizes the importance of Customer, Value and Value Stream, the concept of "Flow" is most critical to Customer Value Delivery. Below, for comparison sake, I've charted the 7 Flows as they're applied to Manufacturing, Healthcare and Healthcare Architects (an example of a Healthcare Service and Supply Provider).

Flow	Manufacturing (original)	Healthcare	Healthcare architecture
1	Raw material	Patients and "Care Givers"	Owners and clients
2	Work-in-process	Care providers	Service providers
3	Finished goods	Pharmaceuticals	Design(s) and plan(s)
4	Operators	Equipment and devices	Machines and tools
5	Machines	Services, materials, supplies	Services, materials, supplies
6	Engineering and processes	Processes and methods	Processes and methods
7	Information and data	Information and data	Information and data

When designing a solution, the design team can take into consideration the 7 Flows and plan accordingly to be sure and address every flow. The 7 Flows provides a frame of reference for resource planning and system wide coverage.

Process problem-solving tools

From my exposure to Quality, Lean, and Six Sigma, I've come to appreciate three specific problem-solving and improvement tools. Each has its own unique features, but all of them follow a scientific method, pattern, and sequence.

- PDCA (or PDSA)
- DMAIIC
- A3

Even before DMAIC (in its popular form) and A3 became frameworks of choice, PDCA (sometimes referred to as PDSA) was a simple, straightforward problem-solving and change method for improvement. All three can be used in unison, each one can be used separately, or two of them can be used in complimentary fashion.

PDCA

PDCA is a commonly known acronym for Plan-Do-Check-Act (Figure 8.1). Many organizations have adopted it as a problem-solving approach. It's a rapid four step model for change. Some organizations use the variant PDSA (Plan-Do-Study-Act). PDCA is often known as "The Deming Cycle" and sometimes referred to as "The Shewhart Cycle." The ASQ (American Society for Quality) recommends its use as follows:

- As a model for continuous improvement.
- When starting a new improvement project.
- When developing a new or improved design of a process, product, or service.
- When defining a repetitive work process.
- When planning data collection and analysis in to verify and prioritize problems or root causes.
- When implementing any change.

PDCA operates as a four-step sequence:

1. Plan. Recognize an opportunity and plan a change.
2. Do. Test the change. Carry out a small-scale study.

P.D.C.A.

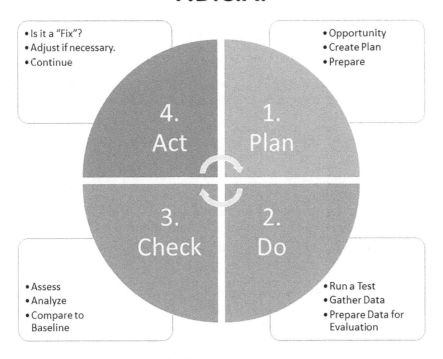

- Is it a "Fix"?
- Adjust if necessary.
- Continue

- Opportunity
- Create Plan
- Prepare

4. Act

1. Plan

3. Check

2. Do

- Assess
- Analyze
- Compare to Baseline

- Run a Test
- Gather Data
- Prepare Data for Evaluation

Figure 8.1 PDCA—Plan Do Check Act.

3. Check. Review the test, analyze the results, and identify what you've learned.
4. Act. Take action based on what you learned in the study step: If the change did not work, go through the cycle again with a different plan. If you were successful, incorporate what you learned from the test into wider changes. Use what you learned to plan new improvements, beginning the cycle again.

Because it's cyclical in nature, it supports ongoing, continual, continuous improvement.

DMAIIC

DMAIC, originally defined and used by Motorola (Bob Galvin, CEO) in the early and mid 1980s to achieve $16 Billion in savings, and later used by Allied Signal (Larry Bossidy, CEO) and GE (Jack Welch, CEO) in the 1990s as a system of management, has demonstrated its capability of solving problems, improving processes, changing an organization, and transforming a culture. Primarily, DMAIC is a data-driven quality strategy used to improve processes. It is an integral part of a Six Sigma initiative, but in general can be implemented as a standalone quality improvement procedure or as part of other process improvement initiatives such as lean.

DMAIC is an acronym for the five phases that make up the process (Figure 8.2):

- Define the problem, improvement activity, opportunity for improvement, the project goals, and customer (internal and external) requirements.
- Measure process performance.
- Analyze the process to determine root causes of variation, poor performance (defects).
- Improve process performance by addressing and eliminating the root causes.
- Implement the process in a standard way so as to enable deployment throughout several sites (This is an IISE add upon realization solutions should be packaged and prepared).
- Control the improved process and future process performance.

The DMAIC process easily lends itself to the project approach to quality improvement encouraged and promoted by Juran. Often, Six Sigma projects take 9–12 months for complete implementation. It is recommended that organizations consider Lean first, then Six Sigma. This provides the tool boxes of Lean and Six Sigma, often referred to as; "Lean Six Sigma."

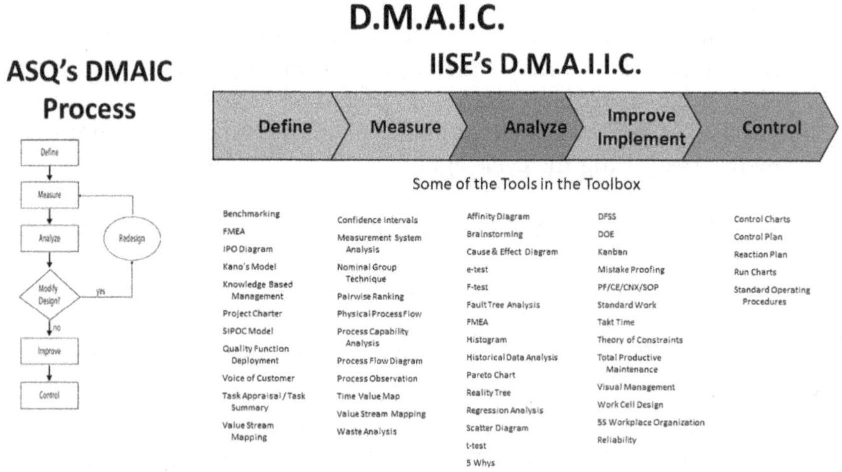

Figure 8.2 ASQ's DMAIC and IISE's DMAIIC for problem-solving.

A3 framework

A3 is a structured problem-solving and continuous improvement approach, first employed at Toyota and typically used by lean manufacturing practitioners. It provides a simple and disciplined approach that systematically leads to problems solved and processes improved. A3 if framed to emphasize problem-solving with structure. It's placed on an ISO –ISO "A3" single sheet of paper, hence the title A3. A3 is has been referred to as SPS, or "Systematic Problem Solving." The process is based on the principles of Edward Deming's PDCA (Plan-Do-Check-Act), and has also been mapped to Six Sigma's DMAIC (Define, Measure, Analyze, Improve, Control).

The A3 approach is divided into a number of steps which can vary. Often, there are eight (8) problem-solving steps mentioned. Some examples of A3 Problem-Solving Steps are (see Figure 8.3 for more):

1. Background: Problem description, Initial Perception (PLAN)
2. Current State: VSM, Process Maps, Baseline Data (PLAN)
3. Goals and Objectives: Point of Cause, Setting Target (PLAN)
4. Analysis: Breakdown of the Problem, Problem Clarification (PLAN)
   - Mapping out for this step can be driven by a set of questions. For example, the "5 Why's" and 5 W's (what, where, when, why, who) and 2H's" (how, how many)
   - Root Cause Analysis, Cause and Effect, Ishikawa
5. Future State and Counter Measures for Containment (PLAN)
6. Implementation and Corrective Actions (DO)

A3 Template

Background	Future State and Counter Measures
- Description of 'What's Going On'	- Future State VSM and Process Maps
- The Reason(s) for Focusing on this Process/Problem	- Potential Actions To Be Taken and Quick Wins
- Context, Content, and Importance	- Counter Measures: First Steps
Current State	Implementation Design and Plan
- The Problem Statement and Definition	- Solution Design and Plan
- The 'As Is' VSM and/or Process Maps	- What, When, Who, Where, How
- The Baseline Data	- Current Status
Goals and Objectives	Business Case and Impact (Optional)
- Target Levels of Performance	- Trend
- Baseline and Desired Outcomes	- Results
- Metrics, Measures, Methods for Performance	- Next Step(s)
Analysis	Follow Up
- Waste Identification and Improvement Opportunities	- Current Status
- Root Cause(s) Typical Tools: Fishbone, 5 Why's	- Next Actions
- Data Diagrams: Pareto, Trends, SPC, etc.	- Recognition of Accomplishment and Achievement

Figure 8.3 A3 template for problem-solving.

7. Review, Follow Up, Status and Progress, Effect Confirmation (CHECK)
8. Communicate and share successful actions and adjust if necessary (ACT)

In the pursuit of Healthcare Affordability, A3 is used for kaizen events, projects and process problem-solving and improvement. It has a similar flow and format to some H&P (Patient History and Physical Examination) templates that are commonly taught in medical school, and used in hospitals, clinics, and other care centers.

People often ask, "Which one is best?" The answer may be, "In some cases, none of them because the organization has adopted its own approach." As an example, some machine and device providers use an "8D" method for problem-solving. Although it has a very comparable flow and content, it uses a different template with unique verbiage. The "8D," short for "Eight Disciplines," has the following components (Note: in this ASQ version there are actually 9):

- D0: Plan—Plan for solving the problem and determine the prerequisites.
- D1: Use a team—Establish a team of people with product/process knowledge.
- D2: Define and describe the problem—Specify the problem by identifying in quantifiable terms the who, what, where, when, why, how, and how many (5W2H) for the problem.
- D3: Develop interim containment plan; implement and verify interim actions—Define and implement containment actions to isolate the problem from any customer.
- D4: Determine, identify, and verify root causes and escape points— Identify all applicable causes that could explain why the problem occurred. Also identify why the problem was not noticed at the time it occurred. All causes shall be verified or proved, not determined by fuzzy brainstorming. One can use 5 Whys and cause and effect diagrams to map causes against the effect or problem identified.
- D5: Choose and verify permanent corrections (PCs) for problem/ nonconformity—Through preproduction programs, quantitatively confirm that the selected correction will resolve the problem for the customer.
- D6: Implement and validate corrective actions—Define and implement the best corrective actions.
- D7: Take preventive measures—Modify the management systems, operation systems, practices, and procedures to prevent recurrence of this and all similar problems.

- D8: Congratulate your team—Recognize the collective efforts of the team. The team needs to be formally thanked by the organization.

However, if an organization is considering using one of the three Problem-Solving Methods showcased here, they can be compared as follows:

A3 element	PDCA	DMAIIC
• Background	Plan	Define
• Current state	Plan	Define
• Goals and objectives	Plan	Measure
• Analysis	Plan	Analyze
• Future state and counter measures	Plan	Improve and implement
• Implementation design and plan	Do	Improve and implement
• Business case and impact (optional)	Do	Improve and implement
• Follow-up	Check/act	Control

Healthcare Affordability depends on people solving the problems of the process, as well as problems in the organization. Affordability cannot be pursued unless problem-solving is pursued, people solving the problems is indoctrinated, and customer value delivery is optimized through process improvement. Included are business problems, organizational structure problems, and performance problems. Complex problem-solving is in the center of the heart of Healthcare Affordability.

Team tooling

The Tuckman Model (An excellent articulation by Peter Scholtes is in his book: "The Team Handbook"[2]) is a model to use for team development. It consists of four stages: forming–storming–norming–performing. This model for group development was first proposed by Bruce Tuckman in 1965, who said that these phases are all necessary and inevitable in order for the team to grow, face up to challenges, tackle problems, find solutions, plan work, and deliver results (below the four phases as described according to Tuckman):

Forming

The team meets and learns about the opportunities and challenges, and then agrees on goals and begins to tackle the tasks. Team members tend to behave quite independently. They may be motivated but are usually relatively uninformed of the issues and objectives of the team. Team members are usually on their best behavior but very focused on themselves. Mature team members begin to model appropriate behavior even at this early phase. The meeting environment also plays an important role to model the initial behaviors of each individual. The major task functions also concern orientation. Members attempt to become oriented to the tasks as well as to one another. Discussion centers around defining the scope of the task, how to approach it, and similar concerns. To grow from this stage to the next, each member must relinquish the comfort of non-threatening topics and risk the possibility of conflict.

Storming

In this stage "...participants form opinions about the character and integrity of the other participants and feel compelled to voice these opinions if they find someone shirking responsibility or attempting to dominate. Sometimes participants question the actions or decision of the leader as the expedition grows harder...." Disagreements and personality clashes must be resolved before the team can progress out of this stage, and so some teams may never emerge from "storming" or re-enter that phase if new challenges or disputes arise. In Tuckman's 1965 paper, only 50% of the studies identified a stage of intragroup conflict, and some of the remaining studies jumped directly from stage 1 to stage 3. Some groups may avoid the phase altogether, but for those who don't, the duration, intensity and destructiveness of the "storms" can be varied. Tolerance of each team member and their differences should be emphasized; without tolerance and patience the team will fail. This phase can become destructive to the team and will lower motivation if allowed to get out of control. Some teams will never develop past this stage; however, disagreements within the team can make members stronger, more versatile, and able to work more effectively as a team. Supervisors of the team during this phase may be more accessible, but tend to remain directive in their guidance of decision-making and professional behavior. The team members will therefore resolve their differences and members will be able to participate with one another more comfortably. The ideal is that they will not feel that they are being judged, and will therefore share their opinions and views. Normally tension, struggle, and sometimes arguments occur. This stage can also be upsetting.

Norming

"Resolved disagreements and personality clashes result in greater intimacy, and a spirit of co-operation emerges." This happens when the team is aware of competition and they share a common goal. In this stage, all team members take the responsibility and have the ambition to work for the success of the team's goals. They start tolerating the whims and fancies of the other team members. They accept others as they are and make an effort to move on. The danger here is that members may be so focused on preventing conflict that they are reluctant to share controversial ideas.

Performing

"With group norms and roles established, group members focus on achieving common goals, often reaching an unexpectedly high level of success." By this time, they are motivated and knowledgeable. The team members are now competent, autonomous, and able to handle the decision-making process without supervision. Dissent is expected and allowed as long as it is channeled through means acceptable to the team.

Supervisors of the team during this phase are almost always participating. The team will make most of the necessary decisions. Even the most high-performing teams will revert to earlier stages in certain circumstances. Many long-standing teams go through these cycles many times as they react to changing circumstances. For example, a change in leadership may cause the team to revert to *storming* as the new people challenge the existing norms and dynamics of the team.

As each stage is applied to Healthcare Affordability (The chart below is from the healthcare research study "Maximizing Team Performance: The Critical Role of the Nurse Leader," presented by Wikipedia[3]) various "Leadership Strategies" need to be deployed, with a variety of "Keys to Success";

Team development stage	Leadership strategies	Keys to success
Forming (setting the stage)	Coordinating behaviors	• Purposefully picking the team • Facilitate team to identify goals • Ensure the team development of a shared mental model
Storming (resolving conflict and tension)	Coaching behaviors	• Act as a resource person to the team • Develop mutual trust • Calm the work environment
Norming & performing (successfully implementing and sustaining projects)	Empowering behaviors	• Get feedback from staff • Allow for the transfer of leadership • Set aside time for planning and engaging the team
Outperforming & adjourning (expanding initiative and integrating new members)	Supporting behaviors	• Allow for flexibility in team roles • Assist in the timing and selection of new member • Create future leadership opportunities

Note: Although this model is not "the only" one to use, it seems to be the one that is most effective with the entire Healthcare Enterprise and all its elements.

Team structural tooling

As discussed in Chapter 4, a dynamic structure comprised of a guidance team, a design team, and several improvement teams should be created for the purpose of making change. The resources on the improvement teams should be engaged in problem finding and problem-solving as a part of their everyday work life. Their efforts should be protected by the guidance team and design team, and their management must be made aware that the work they are doing is part of their job, not in addition to their regular work.

Strategically, if orchestrated correctly, the guidance team can spend several half days over a 4 month period with the design team to create a strategy, a design, and the plan. The plan's horizon can be as short as a year, or as long as 3–5 years. The content of the plan should cover direction, purpose, vision, mission, goals, leadership, management, and team responsibilities. All resources, processes, design, and plan elements should be packed and prepared for the use of the improvement teams and deployment by the design team.

Operationally, the design team is responsible for deploying the design and supporting the efforts of the improvement teams. The design team should be configured to contain; the lead for the effort, respected management personnel who are advocates for the effort, subject matter experts to support the effort and any individual who can provide value for the improvement teams. The prescribed size for this team should be 5–8 with an absolute upper limit of 12.

Lastly, but most importantly, the improvement teams should be put to task to solve problems and improve processes. The list of goals, objectives and targets for the improvement teams should be crafted by the design team, and sanctioned by the guidance team. It is the responsibility of the improvement teams to prioritize their activities and determine the order of actions.

Prioritization tools

Once the team is formed and stormed, during norming, the prioritization of activities arises as a challenge that requires a standard method and methodology. Every individual on the team may have their own method, and individuals may also have differing opinions. Standardizing a priority setting method helps to streamline a team's approach, that in turn, optimizes the team's prioritization effort and sets collaborative and collective points of focus. For prioritization, there are cost/benefit models, risk models, affinity models, and other model options to choose from. Here, I'd like to offer a few models I've found beneficial that may not be common knowledge to most people. In terms of ease, payoff and efficiency, I recommend the following:

- The Ease of Implementation Matrix (Figure 8.4)
- The Payoff Matrix (Figure 8.5)
- The Efficiency Matrix (Figure 8.6)

Each of these matrices can be used by a team in additional to other prioritization methods and models to assist in setting priorities for the solution design, schedule, and implementation.

Ease of implementation matrix

From various research studies, it has been documented that more than 70% of the change and transformation initiatives fail. One of the factors comes from efforts that institute long-term projects for their success, without consideration of short term wins, or sometimes referred to as "low hanging fruit." If, within the initial phases of a program (90–120 days),

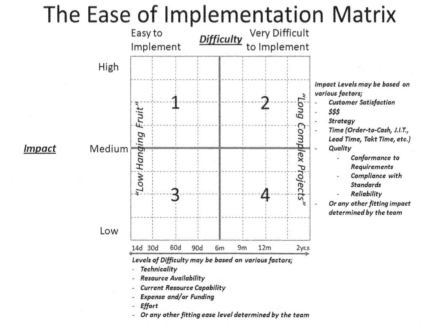

Figure 8.4 Prioritizing solutions using difficulty and impact.

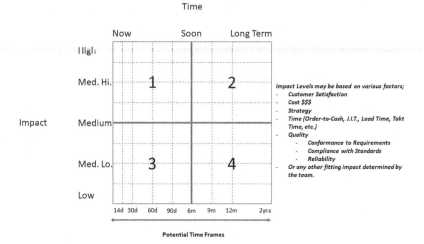

Figure 8.5 Prioritizing solution using time and impact.

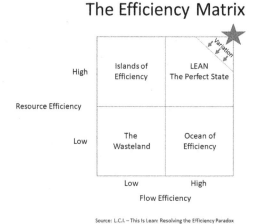

Figure 8.6 Determining lean efficiency using flow efficiency and resource efficiency.

victories and successes are not realized, leadership gets anxious, management gets concerned, and some of the people become complacent. In order for strategic change and transformation initiatives to gain momentum and energy, early quick wins must be realized and recognized to establish enthusiasm. Many of these quick wins should be executed in the first 30 days after the kickoff of the effort.

In order to identify, and prioritize, the activities and events of a program, the Ease of Implementation Matrix can be used (Figure 8.4). It's comprised of two dynamics: Difficulty and Impact. Some activities and kaizens are easy to implement. Other kaizens and projects take longer, and are of greater difficulty. Some events have high impact, while others have low impact. When a team discovers and identifies the activities and events it wishes to pursue, the desired pursuits can be placed on the matrix where they best align in terms of Difficulty and Impact. This endeavor should be done as a team, and every team member should reach agreement on each and every location placement of the planned activities.

For the purpose planning and scheduling, Quadrant 1 should be considered for the first phase of the rollout and deployment of the design. Quadrant 2, should follow, and Quadrant 3 should be considered last (NOTE: Some teams choose to pick "Quick Wins" from Quadrant 3 if there are enough resources to apply.). Eliminate Quadrant 4, or put it "In the Parking Lot" for the time being, with consideration of entertaining those items at a later date. This particular approach assists in the order and scheduling of Quick Wins or Short Term Wins, JDIs ("Just-Do-Its"), Short-Term Projects and Long-Term Projects. By framing the time, the team can determine availability of resources and timeliness of each action.

Payoff matrix

Similar to The Ease of Implementation Matrix, The Payoff Matrix two dimensions of Time and Impact. In order to use this tool, the team places the intended activities on the matrix with consideration of time and impact. The consideration of "impact" may take on a variety of perspectives (e.g., Customer, Cost, Strategy, Time, Quality, etc.). The timeline can be configured to the team's preferences regarding timeframes and time windows. Several Payoff Matrices can be considered to paint various views for prioritization purposes. When comparing the Ease of Implementation results, and the Payoff Matrices results, the team can create a collaborative configuration of what to do when.

The efficiency matrix

Efficiency in Delivering Value to the Customer is about speed and resource capability. The Affordability model addresses this as the relationship between Value and Customer, and Cost and Value. Variation in the speed of delivery and delivery performance, and variation of the quality of resources utilized and applied, together combine to illustrate overall efficiency. I've adopted the "L.C.I.—This Is Lean: Resolving the Efficiency Paradox" approach using The Efficiency Matrix (Figure 8.6)

According to T. Netland, The efficiency matrix says that operations can be optimized along two axes: *flow efficiency* or *resource efficiency*. Flow efficiency is a measure for how fast and undisturbed a customer demand is fulfilled. Resource efficiency is a measure of the extent to which the resources a company possesses are utilized in the processes. A high score on these two different objectives rests on partially competing premises; high resource efficiency benefit from high inventories and long throughput times, whereas the opposite is true for flow efficiency. How to resolve this "efficiency paradox," is exactly what Niklas and Pär want to get at.

The efficiency matrix suggest that a company can have four different states of efficiency; "the wasteland," "ocean of efficiency," "islands of efficiency" and "the perfect state." Reaching the perfect state, a.k.a. Lean, is not easy, and not even strategically right for all companies. *Variation* is lean's biggest enemy. As variation increases, the company is forced to choose between optimizing either flow or resource efficiency. For example, if the company chooses to build inventory, resource efficiency will be improved on the expenses of flow efficiency. Seeking the perfect state— *Lean*—is a strategic choice that is not right for all companies, always." The Lean choice must be part of the Design and Plan for Success to be deployed as a major initiative.

An implementation team standard for kaizens and projects

For team planning and execution, the following seven phases can be used as milestones and guidelines for understanding progress, reporting status, and communicating

1. Project Start
2. Initiation
3. Planning
4. Execution
5. Control
6. Closure
7. Finish

Detailed information for these phases can be found in the Project Management literature.

Step by step

Designing an Affordability implementation takes a team, and as described above, a team of teams. The leader of the effort should have an able-bodied team to support design, scheduling, and deployment. Appendix B illustrates the flow and sequence of activities and events for solution designing and planning. The step-by-step sequence is as follows:

1. Using the output form the "Assess" Phase, have your team analyze the information, data and results to gain a common understanding and comprehension of the requirements and needs.
2. Have the team render a design specifically of how people, process, and performance will all be included in an improvement initiative that creates a better Value Stream, Customer Delivery System for Affordability.
3. From the design, create a plan of; tasks, milestones, schedules, resources.
4. Check for "Completeness" (as determined by the team) of the Design and Plan.
5. The team should finalize the design, plan, and approach for implementation. At this point leadership and senior management should be involved and engaged.
6. The design and plan should be packaged and prepared for implementation and deployment.
7. Preparation should be made for rolling out the design and plan, message communication should commence using the team, top leadership, senior management, and management.

Once the Design and Plan is ready, move forward and enter the "Implementation" stage.

References

1. W. Edwards Deming, 2018, "Point 14 of the 14 Principles", http://asq.org/learn-about-quality/total-quality-management/overview/deming-points.html.
2. Peter Scholtes, "The Team Handbook", 2003, https://www.mindtools.com/pages/article/newLDR_86.htm.
3. Wikipedia, 2018, "Maximizing Team Performance: The Critical Role of the Nurse Leader," https://en.wikipedia.org/wiki/Group_development.
4. LCI, 2017, https://www.leanconstruction.org/lean-i Determining lean efficiency using flow efficiency and resource efficiencyn-action/lean-expert-insight-technology-flow-efficiency-and-a-grass-roots-lean-journey/.

chapter nine

Implement improvements

You've assessed the situation, you've designed and planned the implementation, now's the time to implement. You spent a great deal of time with your team researching the current state and conditions, finding opportunities for improvement, defining challenges to pursue, and developing the schedule for deployment. What should you do to ensure success? Primarily, be sure to have the five components of the "Success Formula" in place:

- Purpose, Vision, Mission, Goals, ... the direction, alignment, motivation, communication.
- Leadership ... top leadership, senior management, management, all aligned and engaged.
- People ... selected, trained, developed, assigned, involved, applied, and aligned.
- Processes and Resources ... means, methods, tools, procedures, support, sources, etc.
- Design and Plan ... the schedule, solution, and program to be deployed and executed.

The factors of the "Success Formula" are not the end-all, be-all, complete list of considerations. For a successful implementation, a comprehensive approach contains all strategy, systems, structure, skills, style and staffing concerns and solutions. An affective approach contains the what, when, where, how, and who in detail.

The overarching intent of the program should concentrate on problem finding, problem-solving, increasing speed, improving quality, and lowering cost, while improving satisfaction for the customers, people, and suppliers throughout the entire stakeholder realm and value stream enterprise. It's about putting people to work to improve processes and increase performance. A suggested supplemental reading for this chapter is: *Affordability: Integrating Value Customer and Cost for Continuous Improvement* Chapter 10, "Process: Work, Work, Work!"

The Healthcare Affordability Roadmap Template (Appendix C) provides a five-phase approach over a 16 month time horizon. It offers recommendations for the sequencing of events, the roles, and responsibilities for the participants, suggested activities for the initiative, and possible actions to take at each phase. Each phase is configured to build off the previous phase, and the last phase assumes that there will be a following phase in the next round of the program. The launch of a Program, like Affordability, can take anywhere from 6 months to 2 years, or even more. It all depends on the current state of the organization and the intent of top leadership and the individuals championing the initiative. One thing is certain, the best probability of success happens with a great deal of pre-planning and planning.

Using Appendix C, I'd like to conduct a walk-thru of each phase and each element, and discuss what is plausible and practical for every timeframe;

Pre-planning, setup, and preparation for the implementation

Even before the designing and planning for deployment occurs, top leadership and the initiative champions, should have in place: a comprehensive definition of the initiative and the program, a consistent message for communication, a select list of individuals targeted for participation, a draft of potential activities, an expected timeframe, a draft of core program principles (purpose, vision, mission, goals, etc.), a process and resource list for participants to use, and a consensus of performance metrics and measures (qualitative and quantitative).

The select list of facilitators, design team members, and guidance team participants should be informed of their upcoming role in the implementation deployment. This may require a collaborative effort on the parts of the top leader and the initiative champion. It requires that the manager of each person on the select list be informed as well.

For the kick-off gathering at the beginning of month one, a purpose-agenda-timeframe must be crafted and communicated for each participant, and their manager, prior to commencement. Every individual must be informed and prepared for; what to expect, how to be prepare, the purpose of the oncoming initiative. Setting the stage for the people involved, and the management impacted, helps in addressing concerns, worries, and "irks."

All information, data, material, and activities should be prepared and ready for month one. The core team of top leadership, champions and assistants should play a personal role in terms of physical, intellectual, emotional and spiritual involvement. In other words, everyone in this group should have had a hand in preparation, involvement of their ideas, engagement of their heart, and full involvement with the team. The commitment of this group is paramount to success.

Initial designing and planning phase

During month one, the core team, the guidance team (the leadership team), the design team (the team operationalizing the events and activities), and the select list of participants will form up to establish a collective understanding of the key principles, philosophies, and intentions of the program. This includes in depth discussion and understanding of; purpose, direction, vision, mission, goals, objectives, processes, resources, metrics, and success measures.

Each team will establish their paradigms, norms, and boundaries. The guidance team will charter themselves to lead, guide, direct, and steer the endeavor. The design team will charter with the guidance team, their responsibility for carrying out the actions and activities necessary for implementation. And, any formed activity or kaizens teams must also determine how they will operate as a team per the influence and facilitation of the design team members.

Month two will focus on training and developing the people chosen to participate in the early stages of deployment. Specific workshops, defined and developed during preplanning, will be held for the various teams to educate and cultivate participants. At this point, program details, elements, and components will be divulged and revealed. Team participants will gain an understanding, as well as provide feedback for the purpose of the identification of any necessary adjustments to roadblocks and barriers to implementation. In addition, additional requirements, needs, wants, and wishes will be sought by design team members and initiative facilitators. At the strategic level, the implementation of the balanced scorecard should take place if had not yet been implemented.

Month three will be the time for adjustment, prioritization of activities, and preparation of any additional materials required for deployment. Guidance Team and the select top leaders will be finishing the second round of their half-day workshops for putting the final touches on the strategic and operational level design and plan. Performance measurement displays for communication should be completed and ready for installation at the various target areas defined.

Month four will culminate the initial phase with finalization of the design and plans at the strategic and operational levels, finish with the preparation for phase 1 activities and kaizens, complete the training and development of those involved in phase 1 actions, and full implementation of the performance measurement and communication system. At this point in time, the readiness for the deployment should be complete.

Implementation deployment phase 1

This phase kicks off deployment with quick wins, strategy deployment, leadership and guidance team engagement, and design team involvement. The guidance team is providing steering and direction for the program, the design team is deploying the initiatives and activities, the people involved are at task with prioritized activities and kaizen events, performance measurement and communication is in action.

The first month of this phase ("month 5") will demonstrate leadership's full support through message communication and participation. The design team will manage the activities and events, and compile several quick wins to generate momentum and motivation. People involved in the quick win kaizens will experience recognition, reward, and celebration of the successes they achieved. The performance measurement system will post and display the victories, and the groups and units involved will experience accomplishment toward the goal.

In the second month, the responsibility of communicating of the message will move down lower into the organization through the senior management level into the management level. The ongoing activities will produce more change and impetus for improvement. More quick wins and short term gains will begin the establishment of the core purpose of problem-solving.

With the final month of this phase, the guidance team should have matured, stabilized, operationalized, proving themselves to be sincere and of high integrity regarding the program. The guidance team at this point should have compiled enough quick wins to establish program viability and remove doubt from any areas of resistance or opposition. The people involved will have accumulated a history achievement, accomplishment, recognition, and celebration. The performance boards should reflect progress and improvement.

Implementation deployment phase 2

During the onset of phase 2, more quick wins should be in motion, as well as some 4–6 month project undertakings. The short term wins, combined with the larger, longer projects, will yield both the "low hanging fruit" successes, and the more complex, more difficult project results. Leadership continues to serve and lead. Management is deeply involved in process improvement and performance improvement. The people are harvesting victories from problem-solving. The scoreboards, at the strategic level and the operational level, display improvement. Process improvements, Lean and Six Sigma efforts, and performance advancement is occurring and beginning "to take hold."

The first month of this phase should start the second wave of momentum. With the four phases, there should be four waves of momentum. This wave is building upon the quick wins only phase one, to quick wins plus longer-term projects in phase 2. Greater momentum, greater energy. Further development of people should also be taking place through coaching and mentoring. Leadership is in the rhythm of status reviews of performance, asking the right questions, listening to the input of the people, and acting upon the requirements and needs of the program.

The second and third month builds upon the groundswell delivering more quick wins and the outcomes of the initial projects being completed. Performance reporting and review rhythms have been inculcated and process improvement has been instilled in the culture. Problem finding, defining and communicating is no longer taboo, but a point of inspiration to all. Problem-solving has become standardized, with faster, better, more affordable infused in the culture.

Implementation deployment phase 3

Six months have passed since the kickoff and initiation of the program. The wins and victories have accumulated, and high performance is being realized. Leadership has gone beyond coaching and mentoring, and is now directly involved in developing new leadership using the Affordability Leadership paradigm. The strategic deployment continues, and after half a year, motivation is on the rise.

Problem-solving, process improvement, and performance monitoring is well entrenched as part of the culture. The longer-term projects are beginning to add to the successes and victories. A large percentage of the organization would now fall into the believer category. There are some pockets of resistance. Attrition of resisters is beginning to occur. The conservative group that once fell in the "middle of the road," is now beginning to migrate to the side of advocacy and innovation for the new heading.

This is the point of critical mass for momentum. The majority of the organization now falls into the believer category. Almost 13 months have passed, and after more than a year, the tide has turned. The first phase of Affordability is almost completely realized.

Implementation deployment phase 4

The leadership team that remains is in full engagement. The guidance team is reviewing and assessing the next step. The design team is preparing for the next round of the program. Adjustments are being made, assessments are being done. And the celebration of victory is at hand.

The first full year has been completed. The beginning of year two is in play. As the culmination of the first round occurs, the design and planning for the second round is initiated. The guidance team and design team may stay the same. New participants may be added, while old participants may change roles. The stability, standardization, and sustainment of the first round should be fulfilled. The time to develop round 2 is apropos.

Affordability Initiative Round 2 ... Level II of the Affordability Program.

This is where I'll culminate my advice. At this point it should be clear what to do next. My philosophy of Outside–Inside, Inside–Outside, Inside–Inside is now realized. My view from the outside, influencing the inside, and my inside perspective, exposing the outside, has culminated in the inside being able to care for the inside. My part is over. Bon Voyage!

chapter ten

Maintain and sustain the results ... continuously improve

The arrival at maintenance and sustainment stage tends to imply culmination and closure, as well as the onset of a static state. In Affordability, this stage only implies we have arrived at the designed and planned future state, that now becomes the current state, requiring that we seek more improvement opportunities and a new future state. It is not static, but rather dynamic. It is not the end, but the new beginning on the continual improvement path. Maintain is but a bridge, establishing the new state of the process and performance, and maintaining stability, standardization and sustainment for the precursor to the next round of improvement. The end of the maintain phase, re-connects with the assess phase, and the cycle renews itself.

Of course we know that performance improvement is an outcome of process improvement. Suggested Pre-Reading; *Affordability: Integrating Value Customer and Cost for Continuous Improvement*, Chapter 11 "Performance: What Is the Score? Are We Winning or Losing?" Attaining performance improvement does not directly imply sustainment of performance improvement will magically take place. If new systems and process are not maintained, the beneficial gains may not be sustained. Systems and processes, especially those that have direct involvement with people, tend migrate back to their original state if not sustained. Sustainment can be accomplished by measuring, monitoring, reviewing, auditing, and focusing on continuous improvement. It's easier said than done. Many organizations enjoy their accomplishments and successes so much, they often choose to move on to bigger and better challenges. Moving on, without a maintenance plan and sustainment conviction, will see that some percentage of their triumphs will experience setbacks. I've heard it often said, "Maintenance and sustainment is the hardest part of continuous improvement."

Sustaining change, according to iSixSigma (a reputable source of knowledge) can be expressed by the acronym SUCCESS: Sponsor, Understand, Commit, Connect, Enable, Support, Sustain. The Sponsor, or champion, should be a top leadership of the organization, willing to break down barriers, provide support, encourage and recognize accomplishment, assist in choosing resources and projects, and get involved when necessary. The term Understand refers to knowing the opportunity, the root cause needed to be resolved for improvement, the process, the people, and the goals. To Connect is to create a collaborative team working in an interdependent manner with a strong bond and culture. To Enable means to provide knowledge, training, development, ability, and opportunity to get involved and engaged. Support is about all the resources, time, and means providing assistance and participating in the activity. Finally, with Sustain, often comes the most difficult task of maintaining the solution and implementation so that the new state does not regress back to the old defective state.

An article written in the Harvard Business Review: "4 Steps to Sustaining Improvement in HealthCare" by Kedar S. Mate and Jeffrey Rakover, November 9, 2016 cites[1];

1. Choose a Pilot Unit within the Organization (Characteristics: Stable and Aligned with intended Goals, and Good Management "Hygiene" meaning "good management practices are already in place.").
2. Start with the Immediate Supervisor at the Point of Care (POC).
3. Use Early Wins to Build Momentum (aka, "Short Term Wins," or "Quick Wins").
4. Motivate Frontline Clinical Managers by Tackling What Irks Them.

Although this seems like a four-step approach to start an initiative, it is true that if you start the initiative in this manner, you will have a good foundation when it comes time for maintenance and sustainment. And, for all the initiatives that follow, starting them in the same manner will shape the condition in such a way that you will have a better likelihood to amply maintain and sustain them.

Once the implementation stage is complete, most of the structure and systems for maintaining and sustaining improvement should be in place. Implementation and maintain throughout a program will tend to overlap. Some areas will enter the maintain stage before other locations. In order to assure success in the maintain stage, several practices and behaviors should be in position. This is a good checklist to use to ensure the Maintain Stage is ready:

1. Leadership and Management has adopted the new philosophy.
2. Performance Measurement has been instituted and is being utilized.
3. Performance Communication is taking place.
4. Performance Improvement using process problem-solving is in place in the organization.
5. Performance Celebration occurs when successes and accomplishments are realized.

1. Leadership and Management

   All leaders and managers throughout the organization set clear direction for the people under their tutelage. The alignment of their group, including the resources within, are interdependent and work collaboratively with the rest of the business. Motivation of people is a priority focusing on areas such as meaning, mastery, and membership. Transparency and clarity occurs through communication in synch with strategy and operational incentives and policies. Finally, execution is realized through achievement of SMART performance goals and objectives.

   Leaders and managers are engaged with their people and involved with their efforts to succeed. They provide support when needed, and they participate in growing and developing the people. Both patience and perseverance are common attributes. They make a concerted effort of cooperate and collaborate with other functional areas and value stream units, and work in partnership with others to remove barriers and roadblocks that create waste and impact performance.

2. Performance Measurement

   Performance measurement is linked and aligned throughout the organization. Strategic, operational and tactical aims, targets, goals, and objectives are consistent from top to bottom. Strategically, a balanced scorecard is in place for focus on customers, business, process, and people. Operationally, process performance is customer centered, with regularly updated and reviewed qualitative and quantitative measures. From the qualitative perspective, attention to customer, people, and suppliers is maintained. Quantitatively some forms of time, quality and cost are used for problem finding, process improvement, and performance communication.

   The performance of every team, unit, department and division measured daily and reviewed by the people in the process, the process owner. On some regular rhythm, management and leadership participates in performance reviews and actions.

3. Performance Communication

   The rhythm of review should be daily for each and every process. Every individual working in a particular should know how the area is performing as a whole. Performance discussions should be held daily at a specific time, for a specific duration, at a specific location. Daily attention to performance keeps workers abreast of the status of the level of execution of the team.

   Performance boards or huddle boards should be in place and maintained daily. These boards should serve as the location for group performance discussions, communication of opportunities for improvement, status updates on problem-solving efforts, and places for recognition and celebration of accomplishments. Leaders and managers can take advantage of such boards for orchestrating waste walks and quality tours.

Performance communication should be conveyed on a corporate scorecard. A best practice approach uses the Andon technique for expressing status of good (green), caution (yellow), and bad (red). Each division can maintain a scorecard linked to the corporate scorecard. Each department or function can maintain a linked scorecard its division, and each team or unit can link their operational and tactical scorecards to their department or function.

4. Performance Improvement

Performance improvement is realized through problem-solving and process improvement. A focus on performance improvement includes people and process. One quick method for checking and acting upon performance opportunities if Plan, Do, Check, Act (PDCA) used as CAPD;

> PDCA (Plan, Do, Check, Act) → CAPD (Check, Act, Plan, Do)
> > Check: Use the Rhythm of Review to Check Status
> > Act: Identify Opportunities, More Waste, and Performance Improvement Challenges
> > Plan: Use Facts Based Problem-Solving to address defective conditions and create a new Plan
> > Do: Execute the new Plan

Performance improvement includes fast adjustments or JDIs (Just Do Its), short term events or Kaizens, and long-term efforts or projects. Using all three approaches simultaneously provides both quick wins and complex solutions to be attained. When the entire organization pursues solutions, collectively, overall performance increases.

5. Performance Celebration

Often overlooked, celebration of success is a critical factor in maintaining and sustaining momentum and motivation. Success celebration does not always imply money or some tangible remuneration for an action or deed. When customer recognizes employees for their work, or respected leaders show appreciation for "a job well done," employee motivation occurs.

Rewards and recognition can be done on an individual basis, a group basis, a site basis and even a corporate basis. Often, heart-felt congratulatory appreciation goes farther than a cash based reward. It has been documented that when peers, or fellow team members, recognize each other, motivation increases. The outcome of motivation is creativity terms of innovation and improvement, and productivity in terms of performance and profitability. The aim is not to celebrate for the sake of celebration, but to motivate to increase a person's sense of worth, meaning, mastery, and membership within the group.

Maintain: the step by step (see Appendix B)

As it is with the three preceding stages of the Affordability Design (i.e., Assess, Design, Implement) and Plan Basic Flow and Framework, the Maintain phase has several salient steps that are critical. Often, organizations will supplement these steps with additional actions that are required for specific conditions or circumstances. The beginning leverages the actions and activities already taking place, and the ending culminates with a loop back to the beginning of the initial Assess phase. Here are the pertinent steps:

1. As the Maintain phase is entered, results from the many accomplishments and successes will be realized, and there will likely be more improvements to pursue. Continue to pursue those improvements prescribed in the design and plan.
2. As additional identified needs arise, make improvements and adjustments accordingly.
3. The rhythm of measuring, monitoring, reviewing, communicating, and improving should continue.
4. Now is a good time to identify and document new opportunities and emerging challenges for the next round of improvement activities.
5. Use the prioritization tools to prioritize the new list of improvement opportunities.
6. Determine the additional gaps and needs that are now realized, begin researching, or at least, prepare a list of specific point to research.
7. Begin the Assess Phase again.

Reference

1. Kedar S. Mate and Jeffrey Rakover, "4 Steps to Sustaining Improvement in HealthCare," Harvard Business Review, November 9, 2016, https://hbr.org/2016/11/4-steps-to-sustaining-improvement-in-health-care

chapter eleven

Enterprise wide responsibilities

Patients, "We want Healthcare Affordability!" and, the care providers cry, "Faster! Better! More Affordable!" and, the enterprise responds, "We deliver Value to Customer requirements, by way of the right resources at an affordable expense, generating revenue at a reasonable and competitive price, that balances the Cost dynamics, that generates growth, profitability and prosperity."

This is not only for Healthcare Provider initiatives but also for the upstream providers within the Healthcare Enterprise. This is what the customer wants, this is what the providers need, and this is the responsibility of all facets of the enterprise. Too much emphasis has been placed on profitability, and profitability alone. An outcome-based focus on revenue, and revenue cycle, is very inward, definitely not outward, nor patient centered. This focus is on system performance and outcome, as well as direction, alignment, staying competitive, moving forward, increasing demand, growing the business, and realizing a long-term vision.

Making decisions by organization outcomes, and outcome alone, is like "driving down the road looking in the rear-view mirror." The image in the mirror has clarity, while the road ahead is unclear. It is likely an unavoidable disaster is looming out there and, constantly looking out outcomes and expecting different results, is a strategy of insanity and demise for sure (Figure 11.1).

Even worse, making decisions from a financial perspective, and a financial perspective alone, puts an organization on a deadly path of doom. The financial view is very important, however, it's an outcome of the customers' demand, customers purchasing of what you deliver, and how much you can keep after all costs and expenses are covered. The road ahead involves customer and marketplace requirements, competitors' offerings, evolving conditions, and best methods for delivering products and/or services. A successful future is achieved through assessing, designing, implementing, maintaining, directing and leading an organization over the roadblocks and barriers, around the pitfalls and hazards, on the journey of Affordability.

Here's a recent example of an organization with its "eye off the ball," and its maneuvering with its view from the rearview mirror. Several deaths occurred at a rehabilitation center in Hollywood, Florida when, after Hurricane Irma past, residents of the center were exposed to extreme

Figure 11.1 Using primarily a financial view is like driving down the road looking in the rearview mirror.

heat and deadly conditions. It was clear that the organization was not patient centered since its overall rating was 1.8 out of 5 (as published by U.S. News and World Report[1]), 1.8 out of 5 (published by Google with 45 Reviews), and a 1.0 out of 5 (Yelp Rating). In fact, the center was located just across the street from a Hollywood Memorial Healthcare System Hospital. The center has been shut down by the State of Florida. It has a slim chance of being reinstated.

The financial perspective is only part of the picture, and always a critical component of the entire picture of every organization. Healthcare Affordability is about delivering the required Value to the Customer, through the funding and full utilization of the resources necessary, offering products and services at a competitive price to generate revenue and profitability, for growing the business and providing jobs. At the Strategic level, success is measured by the Customers, the Business, the Processes, and the People (as adapted from *The Balanced Scorecard*, developed by Robert S. Kaplan and David Norton in 1992[2]). At the Operational and Tactical levels, success is measured by Qualitative means (Metrics of Customer Satisfaction, People Satisfaction, Supplier Satisfaction) and Quantitative means (Metrics of Time, Quality, Cost). The integration of the Strategic and Operational performance measures provide a platform for every level of an organization to determine its success. At every level throughout the organization top leadership, senior management, management and the people is linked via the qualitative measures (Customer and People), and the quantitative measures (Business and Process).

The linkage of the entire Healthcare Enterprise is achieved through flow (see Appendix D). The flow of value delivery, and the flow of patient and care provider requirements. For each entity in the flow, the customers, the business, the process, and the people are all strategic measures for overall performance evaluation. Within each entity, operational and tactical measures provide feedback specifically for monitoring systems and process performance. Downstream delivery is provided according to the upstream requirements flow, and those involved (customer, people, suppliers) combined with what is provided (care, products, services), complete the picture of Healthcare Affordability.

Delivery ← → Requirements

With a customer centered concentration, Affordability is about taking customer requirements, and delivering the value that meets and exceeds those requirements.

Customer Satisfaction

- Time (Speed, On Time)
- Quality (Conforming to Requirements, Complying with Standards, Continuous Improvement)

The relationship of Value and Customer is delivery of what the customer requires, needs, wants, and wishes.

Expenses ← → Resources

The delivery of value requires the use of resources at some cost. The cost of resources comes at an expense that the organization must bear. The best resources often come at an expense that isn't "cheap." There's a good quote on "cheap" that is fitting; "Pretty ain't cheap, and cheap ain't pretty, but everybody wants it to be pretty cheap, and it ain't." Affordability focuses on delivering value from a "cost of" perspective:

Cost of People

- Cost of Methods and Means
- Cost of Products and Services

Affordability also focuses on providing the "Right Resources" for value delivery;

- The Right Resources (People, Methods and Means, Products, and Services)
- At the Right Time
- In the Right Amount

The relationship of Value and Cost provides what is needed to enable value delivery to the customer.

Price ← → Revenue

For balancing the spend, or resource expense, with the revenue, there is a price to be paid by the customer. More often Healthcare Provider

Patient Price is a set range of charges, and payment is frequently handled by Healthcare Insurance, and other payment methods such as Medicare, Medicaid, and Out-of-Pocket. However, for the entire Enterprise, all providers set prices according to several dynamics. Often, prices are set at what the market will bear, yet price setting can be derived by three primary dimensions:

- Customer Examples and Expectations
- Industry Established Price Points and Norms
- Competitor Benchmarks and Comparisons

The relationship of Customer and Cost comes from a price provided that generates the revenue to fund the resources for value delivery. One major challenge in this area comes from the lead time of the revenue cycle, which is measured from the time a service is ordered and provided, until the time the service rendered is paid.

The enterprise flow view

The Enterprise model itself is quite straight forward and simple (again, see Appendix D). The primary supplier entities at the left provide their products and services to the Healthcare Providers in the middle, who in turn, apply their purpose of care application to the patient on the right. The delivery and flow of care, products, and services flows from left to right (a solid line). The requirements from each component and element of the enterprise flow from right to left (a dashed line). This model depicts the relationship between the delivery of value and the requirements from both Patients and Providers. At each entry and exit point, the system can be measured from a strategic and operational perspective. By balancing, leveling, and smoothing the flow, Healthcare Affordability across the entire Enterprise can be achieved.

Healthcare affordability success

Success of Healthcare Affordability is based upon two comparisons: (1) Delivery of Value >= Customer Requirements and (2) Resource Expense of Value Delivery < Customer Revenue. When the delivery of value is greater than or equal to customer requirements, and, the expense of the resources for value delivery are less than customer revenue, the organization is profitable and stays in business. For this to be done, the demands, or commands, for every organization are as follows:

- Satisfy and delight customers by meeting and exceeding requirements, thereby increasing demand and prosperity.

- Deliver Value using excellent resources, and forever seek to increase value and improve the resources.
- Increase speed and quality while lowering cost and expense.
- Decrease the customers' price by decreasing process, product, and service cost.
- Deploy Healthcare Affordability upstream and downstream, and throughout the Enterprise.

The way to achieve Healthcare Affordability is to have the entire Healthcare Enterprise achieve Affordability.

References

1. U.S. News and World Report, https://health.usnews.com/best-nursing-homes/area/fl/rehabilitation-center-at-hollywood-hills-llc-105021.
2. Robert S. Kaplan and David P. Norton. "The Balanced Scorecard: Translating Strategy into Action," 1996, Harvard Business School Press.

chapter twelve

The end ... start now

So you'd like to start now. At this point, you may want to know:

- Where do I start?
- What do I do?
- How do I do it?

First of all know this; 70% of Organization Transformations fail. The reasons for failure often occur because of a bad purpose, a bad start, a bad design, a bad plan, and an inept leadership team. So, if you were a conservative person, who bets on the favorites, you may choose to not proceed. However, if you do go forward, with a purpose, leadership will be your key success factor. Assess, design, and implement an initiative program of Affordability.

The "Where"

For the best probability of success, be sure and cover all organization levels: strategic, operational, and tactical. Start with a collaborative team design and plan created by advocates and believers that will serve to deploy the program. Start by lining up leadership and people that are innovators and supporters of the initiative. Start by setting direction, aligning resources, motivating the people, communicating the message, and executing the plan. Start by putting in place resources and assuring that the people can be provided time to solve problems and improve processes. Start with the harvesting of low hanging fruit and quick wins. Start by celebrating, recognizing, and rewarding success.

Strategically, start with Leadership and Strategy. For the best results, you must know, going in to deployment, who will lead the efforts, in what direction they will be expected to go, how they will align the resources, how they will communicate the message, and what design and plan will they deploy.

The leadership team should have been selected on the basis of their support and advocacy for Affordability, as well as their influence within the organization. In other words, choose respected leaders that fully embrace the tenets of Affordability.

Here are various sets of steps to consider:

Defining the Strategy—The 14 General Steps and Order of Events at the Strategic Level are (Note: Also be sure and review the Affordability Planning and Initial Deployment Timeline template for transformation, and the Healthcare Affordability Algorithm in Chapter 3):

1. Define the Organization's true Value and Purpose, and the Peoples' Values and Principles.
2. Determine that Powerful Leadership Coalition you will engage and employ.
3. The Vision, Mission, and Goals should be clear and communicated by the Leadership Team.
4. Understand the Customers' and Industry's Requirements, Wants, Needs, and Wishes.
5. Map and Document the Value Stream(s) and Processes.
6. Declare the Value Added and Value-Added Support Functions.
7. Define the Value Delivery and Flow.
8. Identify the Waste and the Opportunities for Improvement.
9. Enact a Strategy/Systems/Structure using a Guidance Team, a Design Team with action Teams.
10. Enable the People to Act and Succeed
11. Put the People to work eliminating waste, solving the problems, and improving performance.
12. Regularly review performance and continue to pursue improvement. ("CAPD")
13. Leverage the Successes and Create more "Wins" and Successes.
14. Institutionalize the changes and pursue transformation.

Setting the Direction

1. Discovery, determine, define Purpose.
2. Establish a Sense of Urgency (We want to … We must … We will).
3. With the Leadership Team (Guidance Team) create a collaborative and shared Vision.
4. Define the Mission.
5. Define the Goals.
6. Craft the Message and Communicate the Message.

Aligning the Resources

1. Establish the Guidance Team.
2. Establish the Design Team.
3. Determine the Right Resources to get involved.
4. Determine the Right Activities and Projects to pursue.

5. Determine the Right Tools and Methods for use.
6. Develop the capabilities of the people in order to enable them to act and institute action.
7. Be bold and promise big.

Motivating the People

1. Communicate the new Purpose, Vision, Mission, Goals, Select Participants.
2. Communicate the Message.
3. Communicate the Design.
4. Communicate the Plan.
5. Accomplish and Celebrate Quick Wins (wins in the first 30 days).
6. Recognize People for what they achieve.
7. Reward those who deliver results.
8. Celebrate Improved Performance.
9. Celebrate, celebrate, celebrate.

Improving the Processes

1. Use select people.
2. Develop problem-solving capabilities in those people.
3. Invoke problem finding and opportunity discovery.
4. Put the people to work solving the problems.
5. Review status and accomplishment.
6. Celebrate Success.

Communicating the Design and Plan

1. Craft the design and plan.
2. Institute the "5 Times Rule" for communicating the message. Communicate the message at least five times, using five different methods and media, at five different times. This of course will end up with more than five times
3. Communicate through quick wins, activities, events, and projects.
4. Communicate through regular review and rhythm of performance assessments.
5. Have top leadership communicate the plan and design.
6. Have Senior Management communicate the plan and design.
7. Have Management communicate the plan and design.
8. In addition, have everyone add the main message and theme of the program to their communication.
9. Celebrate Wins to communicate the successes of the design and plan.

Any, or all, of the sets of steps above can be considered, and may serve to be beneficial when thinking about beginning an Affordability journey. The place to start should be determined by the guidance team, in conjunction with the design team and champions.

The "What"

As for the "What do I do?," a sense of urgency is always a good platform from which to springboard an initiative. This urgency can be founded from a crisis, or an impossible dream, or both. A crisis in the marketplace, or a crisis in the organization. A dream to take the institution to a world class level, or a dream of serving the area with the best healthcare available. The basic "what" addresses purpose.

What you should do as a leader is: set direction, align the resources, motivate people, communicate the message, and execute the plan to implement the design. Your behavior should be shaped such that you; model the way, inspire a shared vision, challenge the process, enable others to act, and encourage the heart and soul of everyone. Through your actions and behaviors, the people within the organization will understand what to do and how to do it.

A critical component of the "what" is the focus on people solving problems in the process to improve performance. This what is: Motivate People → Improve Processes → Increase Performance (Appendix A). What must be done is: (1) Assess, (2) Design, (3) Implement, (4) Maintain (Appendix B). Much of the "what" should be scheduled and planned for around a year or more (Appendix C). From the enterprise perspective, the "what" includes all of the providers of care, products and services (Appendix D).

In the end, it's about increasing performance through improving processes. By measurement, methods of monitoring and reviewing, and techniques of use and utilization, the "what" can be actualized.

1. Establish the System and the Baselines for measuring and monitoring performance
2. Set SMART Goals
   - S Specific (simple, sensible, significant)
   - M Measurable (meaningful, motivating, metric based)
   - A Achievable (attainable, agreed, authentic)
   - R Relevant (reasonable, realistic, results based)
   - T Timely (time based, time bound, time sensitive)
3. Use the system throughout the strategic, operational, and tactical realms.
4. Accomplish and achieve performance improvement.
5. Celebrate!

The "How"

After reading this book, you would guess, the order of importance, and the logical sequence for implementation is: People → Process → Performance. You might also think that Process may be first. It is clear that Process is where the improvement will happen. However, in order to change the process, the people have to be prepared for changing the process. Performance is critical, but, the baseline measures of performance "are what they are," the current state. Performance change will be recognized when the process changes. First and foremost, start with yourself and your leadership team. Affordability happens best when leadership is directing the charge steering the way forward. Using the four cornerstones images for the foundation (i.e., Healthcare Affordability— Appendix A, The Healthcare Affordability Algorithm—Appendix B, The Affordability Planning and Initial Deployment Timeline—Appendix C, The Enterprise Flow Diagram—Appendix D), and the Healthcare Affordability tools and techniques presented in this book, supported by the original Affordability text (Affordability: Integrating Value, Customer and Cost for Continuous Improvement), your program for Healthcare Affordability can be crafted.

The strategic view can be fashioned using the first four chapters and Appendices A, B, and C. The operational and tactical initiatives and be produced using Chapters 5 and 6. The program design and plan can be constructed using Chapters 7–10. And finally, the rest of the groundwork, footings and underpinnings of the framework can be completed using Chapters 11 and 12, and Appendix D. All that is left, before starting, with the where and the what, is to determine the when, the who, and the why.

The when, the who, and the why

As for the "When" and the "Who" that's up to you. My personal suggestion is just do it, do it now. There may be a better time in the near future, but if you consider the near future to be more than a year away, that's too late. Don't put it off.

The who is you. You and a powerful leadership coalition of your choosing that serves as a guidance team to direct, align, and motivate. A collaborative team of believers to serve as a design team for designing the approach, creating the plan, and instituting. And, the advocates in the workforce ready to be unleashed on problem-solving, process improvement, increased performance, greater profitability, and ultimate prosperity.

As for the "Why" … if you don't know the "Why" by now, don't start.

Here are some Observations of Successful Affordability Organizations:

- Top Leadership living the Philosophy ... where Leaders are Involved and Engaged.
- There's a "Want To" and a "Need To" attitude, not just a "We Have To" outlook.
- Teamwork and Cross Functional Collaboration is apparent.
- Development of People is accomplished through Internal Training & Development Programs.
- The Culture is visibly Changed and Transformed from what it once was.
- The organization has a practice of a "willingness to share best practices with others."
- Key Partnerships exist with Providers that fit the Purpose.
- People Solve the Problems and Challenges in the workplace as part of "what we do."
- The Organization Learns and seeks to Learn new and better ways.
- The Organization knows its past, present, and future, and how it's moving forward.
- Performance is measured: Process—Leading Measures, Outcomes—Lagging Measures.
- Communication is apparent and visible.
- Successes and Positive Results are Celebrated.

There is one characteristic that transcends across all organizations that accomplish Affordability: Their solution was created from within, using best practices, without copying nor replicating a particular approach. No organization seems to hold the "enchanted formula" for Affordability. There is no silver bullet, no magic wand, and certainly no free lunch. The best solutions are developed by motivated people, using proven process improvement methods, techniques and tools, and increasing performance by way of leading people and managing processes. Good Luck!

Appendix A

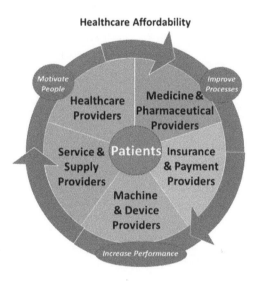

Figure A.1 The Path to Healthcare Affordability.

Appendix B

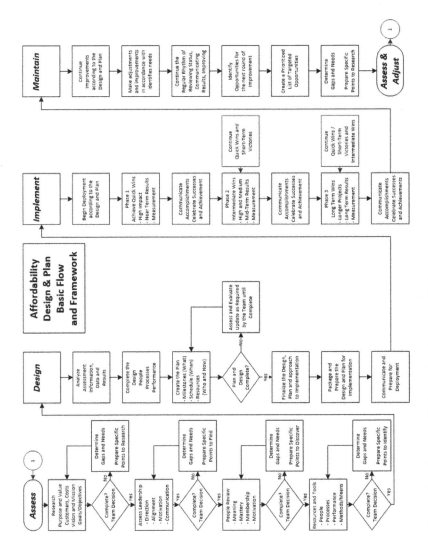

Figure B.1 Assess—Design—Implement—Maintain.

Appendix C

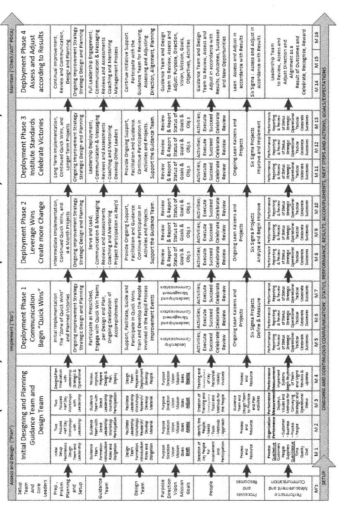

Figure C.1 A-16 Month Template to begin the Pursuit of Healthcare Affordability.

Appendix D

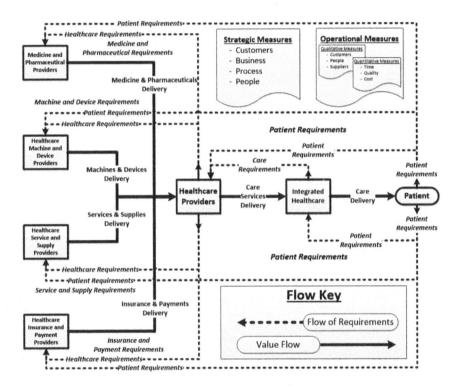

Figure D.1 The Healthcare Enterprise Flow Diagram.

Index

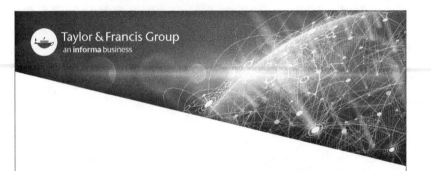